Changing Identifications and Alliances in North-East Africa

Integration and Conflict Studies
Max Planck Institute for Social Anthropology, Halle/Saale

Series Editor: Günther Schlee, Director at the Max Planck Institute for Social Anthropology

Editorial Board: Brian Donahoe (Max Planck Institute for Social Anthropology), John Eidson (Max Planck Institute for Social Anthropology), Peter Finke (University of Zurich), Joachim Görlich (Max Planck Institute for Social Anthropology), Jacqueline Knörr (Max Planck Institute for Social Anthropology), Bettina Mann (Max Planck Institute for Social Anthropology), Stephen Reyna (University of Manchester)

Assisted by: Cornelia Schnepel and Viktoria Zeng (Max Planck Institute for Social Anthropology)

Volume 1
How Enemies are Made: Towards a Theory of Ethnic and Religious Conflicts
Günther Schlee

Volume 2
Changing Identifications and Alliances in North-East Africa
Vol. I: Ethiopia and Kenya
Edited by Günther Schlee and Elizabeth E. Watson

Volume 3
Changing Identifications and Alliances in North-East Africa
Vol. II: Sudan, Uganda and the Ethiopia-Sudan Borderlands
Edited by Günther Schlee and Elizabeth E. Watson

Volume 4
Playing Different Games: The Paradox of Anywaa and Nuer Identification Strategies in the Gambella Region, Ethiopia
Dereje Feyissa

Volume 5
Who Owns the Stock? Collective and Multiple Property Rights in Animals
Edited by Anatoly M. Khazanov and Günther Schlee

Volume 6
Irish/ness Is All Around Us: Language Revivalism and the Culture of Ethnic Identity in Northern Ireland
Olaf Zenker

Changing Identifications and Alliances in North-East Africa

Volume II: Sudan, Uganda and the Ethiopia–Sudan Borderlands

Edited by Günther Schlee
and Elizabeth E. Watson

berghahn
NEW YORK · OXFORD
www.berghahnbooks.com

First published in 2009 by
Berghahn Books
www.berghahnbooks.com

Library of Congress Cataloging-in-Publication Data
Changing identifications and alliances in North-East Africa / edited by Günther
Schlee and Elizabeth E. Watson.
    p. cm. -- (Integration and conflict studies : v. 3)
Includes bibliographical references and index.
ISBN 978-1-84545-604-7 (hardback) ISBN 978-1-84545-963-5 (institutional
ebook) ISBN 978-1-78238-331-4 (paperback) ISBN 978-1-78238-332-1 (retail
ebook)
    1. Group identity--Africa, Northeast. 2. Ethnicity--Africa, Northeast. 3. Africa,
Northeast--Ethnic relations. 4. Africa, Northeast--Social conditions. I. Schlee,
Günther. II. Watson, Elizabeth E., 1968-
    DT367.42.C533 2009
    305.8009676

                                2009047676

British Library Cataloguing in Publication Data
A catalogue record for this book is available from the British Library

            ISBN: 978-1-78238-331-4 paperback
            ISBN: 978-1-78238-332-1 retail ebook

Contents

Part IV Displacement, Refuge and Identification

List of Maps, Plates, Figures and Tables

Maps

Plate

Figures

Tables

Appendices

List of Abbreviations

AK-47	Kalashnikov Automatic Weapon
AMDF	Arba Minch Development Farm
ASO	Anywaa Survival Organization
CPA	Comprehensive Peace Agreement
DC	District Commissioner
DFG	Deutsche Forschungsgemeinschaft
DSC	District Security Committee
EECMY	Ethiopian Evangelical Church Mekana Yesus
EPLF	Eritrean People's Liberation Front
EPRDF	Ethiopian People's Revolutionary Democratic Front
EPRP	Ethiopian People's Revolutionary Party
GOSS	Government of South Sudan
GPDUP	Gambela People's Democratic Unity Party
GPLF	Gambela People's Liberation Front
GPLM	Gambela People's Liberation Movement
HIPC	Highly Indebted Poor Country
IRC	International Rescue Committee
JEM	Justice and Equality Movement
KPDO	Konso People's Democratic Organization
KY	Kabaka Yekka
LRA	Lord's Resistance Army
LWF/DWS	Lutheran World Federation/Department for World Service
MEISON	All-Ethiopia Socialist Movement
NCA	Norwegian Church Aid
NFD	Northern Frontier District
NGO	Non-governmental Organization
NIF	National Islamic Front
NRA	National Resistance Army
NSCC	New Sudan Council of Churches
OAGs	Other Armed Groups
OLF	Oromo Liberation Front
OLS	Operation Lifeline Sudan
OPDO	Oromo People's Democratic Organization
PCOS	Presbyterian Church of Sudan
PRS	Proto-Rendille-Somali
RRA	Rahanweyn Resistance Army
SAD	Sudan Archive, Durham University
SAF	Sudan Armed Forces
SCC	Council of Churches in Sudan
SIM	Sudan Interior Mission

SNNPR	Southern Nations, Nationalities, and People's Region
SPCM	Swedish Philadelphia Church Mission
SPDF	Sudan People's Defence Force
SPLA	Sudan People's Liberation Army
SPLM	Sudanese People's Liberation Movement
SSDF	South Sudan Defence Force
SSIM	Southern Sudan Independence Army
SSLM	South Sudan Liberation Movement
SSUA	South Sudan United Army
TPLF	Tigrean People's Liberation Front
TTI	Teacher Training Institute
UDSF	United Democratic Salvation Front
UNHCR	United Nations High Commissioner for Refugees
UNLA	Ugandan National Liberation Army
UPC	Ugandan People's Congress
UPDF	Uganda People's Defence Force
WFP	World Food Programme
WV	World Vision

Introduction

Elizabeth E. Watson and Günther Schlee

'Who belongs to whom and why' is an enduring question for social science, and for contemporary policymakers who are faced with – and sometimes attempt to manipulate – the processes of collective identification. Decades of academic argument about whether or not forms of collective identification such as ethnicity will disappear as a result of the march of history and modernization have been replaced by an acceptance that forms of collective identification change over time but are as significant as ever for their influence on politics and individual livelihoods (see Turton 1997, for review), shaping the way resources are claimed and distributed (Bayart 1993). In North-East Africa, identifications are central to the ways in which similarity to and difference from others are understood and constructed, and they are connected to processes of inclusion and exclusion, to degrees of openness or closedness to others, qualities that are sometimes described as cosmopolitanism or parochialism. The boundaries that designate relations of self to other result, at least in part, in tolerance or prejudice, legitimizing peaceful or conflictual relations.

Forms of identification influence perceptions of what is expected and what is possible, and they shape individual subjectivities and experiences. Pre-existing social givens govern these processes, but they are also the products of particular histories and result from interactions between processes at local and regional, national and global scales. At the smallest scale, identifications are enacted through and on the body, in dress, style of walking or patterns of scarification. They are produced and performed creatively in relations within and between different groups, through the telling of oral histories and through rituals, songs, dances, interactions and exchanges. They are also produced through marriage, bond friendship and conflict. At the regional, national and international scale, identifications are shaped by exposure to new technologies, ideas and cultural styles and by the structures of formal and informal policies and politics. These govern who gains access to what and when, who is preferred in networks of support and patronage, who is invested in, who is given aid, who benefits and who is excluded from those benefits. It is the coming together of the personal and physical, bodily and relational, on the one hand, with the regional, national and international politics that distribute resources, on the other, that gives particular force to these forms of identification, and makes them powerful, even, at times, explosive.

In thinking about identities here, the division that is commonly made between 'primordial' and 'instrumental' approaches has not been drawn on extensively, for two reasons. First, approaches that see identities as either primordial age-old historical artefacts with strong emotional power or instrumental forms of organization that have developed in response to particular conditions and opportunities often fail to appreciate the way in which identities are formed over time, have porous boundaries and are experienced as highly emotive phenomena. There are grounds for taking elements of each approach (see Spear 1993; Dereje Feyissa, this volume). As Turton writes: 'the analysis of ethnicity must therefore take account of both its "instrumental" or material aspects and its "primordial" or cultural aspects, since its very effectiveness, as a means of advancing group interests, depends on it being seen as "primordial" by those who make claims in its name' (1997: 82). Secondly, approaches to identity and alliance formation that focus on a primordial versus an instrumental approach start and finish with ethnic identity. This analytical lens serves to reinforce ethnicity as the most important form of identity and scale of investigation. Many of the chapters in this volume study aspects of ethnic identity, but they also examine other forms of identity that exist within and across the ethnic.

Many of the individual chapters take their own approach to identities and alliances, but as editors and for the non-specialist reader we have developed the following general framework. The approach to identities and alliances focuses on the interface between structure and agency. Indigenous structures that define identity include forms of descent, clan and moiety systems, marriage patterns, belief systems, livelihood practices and language. These in turn are shaped by the changing processes and contexts such as economic or environmental opportunities and challenges, new religions, government policies and practices and the arrival of new technologies, including weaponry.

But these endogenous and exogenous structures do not combine on their own to influence identities: they have force and impact because of the way in which they are accepted, appropriated, reacted to or resisted by people. Individuals and groups act creatively and give these structures meaning, in processes through which individuals and groups articulate identities and position themselves (Hall 1990). In the introduction to Volume I, Schlee explores how this articulation can be carried out differently, for example, by defining identities of wider or narrower scope, by switching between different facets and ways of defining identity. In different situations it may be advantageous for a group or individual to define themselves in a particular way.

Individual agency and group interests must be taken into account when considering identity, but we must remain cautious about ascribing selfishness or too much strategy and assuming that all identity-related actions are motivated by the desire to promote personal interests and positions. In addition, individuals and groups are not, by any means, wholly free to switch identities or to change meanings as they wish. Some forms of identity and some alliances are more plausible than others. History matters, in terms of with whom it is possible to form alliances; previous forms of identification have a strong influence on what is deemed acceptable; cultural values attached to war or to peace, to environments and territories, influence the likelihood and nature of conflict and peace.

Where war and peace are concerned, the essays in this volume show that there can be no single causal explanation. On the one hand, inter-group conflict can be seen as more likely in situations where the collective identifications have become more salient and more ethno-nationalistic (to use Turton's 1997 phrases). Ethno-nationalism can be understood here to refer to a strong sense of difference and a lack of tolerance for sharing territory and resources. When a strong sense of identity is combined with competitive and negative stereotyping of others, conflict is more common. In some of the chapters that follow, it is clear that the structures that have emerged in recent decades have encouraged more salient and possibly more rigid forms of group identity to emerge. These structures include the way refugee camps are organized and the impact of national boundaries, state policies and religious institutions. On the other hand, some of these essays and other scholarly accounts have also shown that conflict itself can also be a cultural means for re-emphasizing boundaries between groups and for constructing identities (Fukui and Markakis 1994).

These theoretical issues are set out more extensively in the theoretical introduction by Schlee at the beginning of Volume I. The chapters in this volume explore these processes through case studies from Sudan, Uganda and the Ethiopia-Sudan borderlands (see Map I.1). Volume I covers case studies from Ethiopia and northern Kenya. The region shown in Map I.1 is vast and varying, but forms of collective identification and alliance construction have been strong and have had lasting consequences for lives and livelihoods. They are prominent forms of organization through which individuals and groups understand and organize their worlds and mobilize and engage others in social lives and political agendas.

In southern Sudan and northern Uganda, processes of identification and alliance are dominated by experiences of conflict and raiding. Southern Sudan was the crucible of theories of segmentary lineages put forward by E.E. Evans-Pritchard to explain shifting patterns of alliance and conflict. In recent years, the reasons for and nature of the conflict and associated experiences may have changed, but raiding and violence are still commonplace and the shifting patterns of collective identifications are kaleidoscopic. The chapters presented in this volume show that Evans-Pritchard's ideas are by no means redundant. The connection between the local and the international is also particularly apparent here, as local alliances and conflicts are influenced, at least in part, by shifts in international geopolitics and/or by the international trade in small arms or by the actions of multinational oil companies.

Direct experiences of conflict may dominate many people's lives, but more personalized and everyday matters are still significant to individuals and also play a role in creating group alliances and divisions. Even the seemingly trivial matters that make up forms of popular culture are significant to senses of self, to orientation, to a sense of place in the world and to possible options and outcomes. The areas further to the north of Sudan have been much less the subject of classic anthropological texts, but the processes related to identifications there are no less interesting. The chapters that explore this region highlight the role of structures such as marriage and international boundaries and the development of new technologies in constructing and changing identifications and alliance patterns. Also included in this volume are chapters on groups living in and around Gambela town on the Ethiopian side of the border. These areas have been influenced as much by the politics of southern Sudan and the Sudanese state as they have by the politics of the Ethiopian state; the chapters show the way the fates, histories and impacts of the different political entities in the region are intertwined.

Map I.1 The Horn of Africa and the approximate location of the people who feature in these volumes

The chapters demonstrate that explanations for the shifting patterns of identifications are to be found in historically informed studies that take into account the roles of charismatic leaders as much as those of their followers, and the roles of women as much those of men, and that studies need to explore equally economic, political, cultural, emotional and personal motivations. Collective identifications persist as the product of the interrelations between culture, history, politics, biology, technology and perceived interests. They are also shaped by the geopolitical context: identifications shift according to the impact of conflicts at national and international levels or to the impact of international and other boundaries, and identifications change when people move or are moved into the liminal spaces between them. Understanding the intersections of these different processes is the task here, where each place and group of people(s) must be understood in terms of their local as well as their national and international context.

Many other studies of this region construct general overviews of the political and economic processes that unfold. Journalistic accounts focus on the 'problems' of these areas: they cover the conflicts or food shortages, but they tend to rely on simplistic and stereotyped explanations and reveal little about how and why those conflicts and food shortages arise or how they are experienced. The chapters here provide more in-depth information on this area, and they show how the nature of local experiences and reasons for conflicts are often related to the politics of identifications. These chapters also explore how those identifications are experienced and viewed emically. Some of the chapters here look at conflict and the consequence of conflict, but they also explore other, more mundane processes, putting the 'problems' of the region in their wider context. The chapters focus on the consequences of these processes for different groups and individuals on the ground. Although the understanding of particular contexts is shown to be vital, wider theoretical insights can also be derived that bring critical insights into the complex processes of identification and alliance formation.

These chapters are authored by a diverse group of scholars, in terms of age, nationality and discipline, each of whom has carried out significant fieldwork among the people discussed. Their chapters were written in response to a call to examine the shifting identifications and alliances in the region. Their choice to focus on a particular aspect of identification was driven partly by the empirical situation and partly by their own interests and disciplinary orientations. Empirical knowledge and theoretical insights can be drawn from the individual chapters, but it is also hoped that the chapters will intersect with each other to produce a portrait of a region and an understanding of the way processes shape identifications, in the past, present and further in the future.

Later in this introduction we give a brief summary of the chapters that follow and the themes they address. The chapters in Volume II are arranged partly by geographical region and partly by theme. Most of them relate to people in Sudan, and one relates to people in Uganda. Several discuss people in the south-west areas of Ethiopia, around Gambela town, but they are included in this volume because many of the people discussed are Sudanese refugees or their lives have been heavily influenced by the unstable processes taking place in Sudan. One chapter is set in Kenya, in Kakuma refugee camp: this also explores the experience of a group of people from Sudan, in this case the Pari.

As in Volume I, we first provide a short summary of key events in the political histories of Sudan and Uganda, with the aim of making this collection accessible to the non-regional specialist reader, and to prevent each chapter from having to repeat the same background information. Anyone wishing for basic information on the events and histories of Kenya and Ethiopia should turn to the Introduction in Volume I, where these countries' histories are examined more fully. Anyone already familiar with the general history of Sudan and Uganda may wish to go straight to the chapter summaries, and from there to the chapters themselves.

Political Histories

Sudan

Most of the chapters in this volume relate to people who are situated in what is present-day Sudan. A short summary of the history of Sudan, derived from sources such as de Waal (1997), Holt and Daly (2000) and Johnson (2003), reveals astonishing levels of instability and political factionalism. A constant theme is the tension and conflict between north and south, core and periphery, and the different levels of investment and entitlement between them. The core, with its centralized kingdoms, centres of colonial power and theocratic state, has continued to dominate the more acephalous periphery, which served mainly in earlier years as a source of slaves and ivory. The governments of the core have been characterized by interpersonal political rivalries that have played out in support of different factions in the periphery or in support of 'tribal militias' who have been encouraged to wage war. Instability, violence and famine in the periphery have fed back into further factional struggles at the core. Competition for oil has now fed into these cycles of violence and instability, as has the impact of international alliances and resource flows, which have also changed dramatically as a result of fallout from 11 September 2001.

The chapters in this volume that relate to the northern part of Sudan, the Nile and Blue Nile regions, discuss people who are often described as Arabicized or Islamicized. The literature describes these areas and many others, including Darfur, as once being the home to African kingdoms and royal clans who rivalled each other for power, territory and subjects, from whom they collected tribute. Chief among these, and mentioned in several of this volume's chapters, is the Funj kingdom, thought to have been established early in the sixteenth century and based at Sennar on the Blue Nile. The Funj territory was extensive. It relied on local notables to manage trade, to protect its subjects and to collect tribute. The rise of the Funj kingdom was accompanied by its conversion to Islam, as a result of the influence of Arab immigrants from the north.

The Funj kingdom, or sultanate, was on the wane before the Egyptians under the rule of Muhammed ʿAli Pasha turned their attentions to the south. In the early nineteenth century, Egypt was under the rule of the Ottoman empire, although Muhammed ʿAli Pasha had extensive autonomy. He executed a harsh and brutal conquest of the Sudanese territories, driven by the need to control what he saw as a political threat and encouraged by the possibility of access to reserves of slaves and gold. The Egyptians' superior weaponry enabled them to overpower local kingdoms and to tax their subjects. Those who refused or were unable to pay had their children, slaves or cattle seized.

By the mid-nineteenth century the Egyptians had begun to penetrate further into the south, in search of more resources. The trade in ivory was highly lucrative, and many European traders and other adventurous entrepreneurs were involved in travelling to the south to obtain it, sometimes by force, from local people. Slaves were traded or captured, either instead of ivory that could not be found or to be used to transport the ivory. These traders were based in fortified settlements, or *zaraaʾib* (singular: *zariiba*; colloquially: *zariibaat*; often transcribed *zariba*), described by Johnson in his chapter.

The Egyptian regime suffered problems at home towards the end of the nineteenth century. Khedive Ismail, then ruler of Egypt, overextended his regime in struggles in Syria and Palestine. He incurred massive debts, which gave foreign powers stronger influence over Egyptian affairs. By then the 'Turkish' (Egyptian) military had already incorporated many European officers, among them Gordon Pasha (Charles G. Gordon), Emin Pasha (Eduard Schnitzer), Salatiin Pasha (Rudolf von Slatin), Romolo Gessi and many others, and the difference between the Egyptian rule and the rule of European colonial powers established elsewhere in the same period became blurred. At the same time, Khedive Ismail was a more cosmopolitan ruler than his predecessors, and he had declared that he was against the slave trade. He was not able, therefore, to make use of one of the main resources of Sudan as his predecessors had done, and this may have meant that his interest in the region was less strong. The Egyptians' control over the large territories of Sudan became increasingly tenuous, and a series of revolts took place.

The revolts resulted in the establishment of the Mahdist (or Ansar) state (1881–98). It controlled and administered large swathes of current-day Sudan, extending even into what is now western Ethiopia. The Mahdist state was a theocracy founded by Muhammed Ahmad ᶜAbdallah, heralded as the 'Expected Mahdi'. His coming was thought to presage the beginning of a just and equal society brought about through the good practice of Islam and the Mahdi's guidance, and he claimed to gain his authority and judgement directly from God. His strong religious ideology and his charisma brought him the commitment of many followers, although many were also attracted by the way the Mahdists had routed the foreign rulers.

The Mahdists instigated a system of taxation of their own, and continued to wage war against other groups in an attempt to bring all under their holy suzerainty. They fought with the troops of Emperor Yohannes in Ethiopia, and occupied the territories around Assosa and sacked Gondar. When the Mahdi died, his successor's rule lost some of the religious and nationalist fervour that had given the Mahdists momentum and support. The Mahdists continued to wage holy war, but the regime was subject to internal rivalries between different factions.

The Mahdist period was ended by the British campaign under Sir Herbert Kitchener. British interest in Sudan had been triggered by the 'scramble for Africa', the competition for control over territory, resources, labour and markets in the continent. The British in Egypt were also concerned to neutralize a threat they perceived from what Lord Cromer in 1908 referred to as 'the Dervish hordes at Omdurman'. In 1899, the Egyptians and British agreed to exercise joint control over Sudan, described as the Anglo-Egyptian Condominium. In practice, the British had great autonomy, although their new subjects discriminated little between the nationalities of their rulers, and referred to them all as 'Turks'.

As elsewhere, the British set up a system of indirect rule, organized by 'tribe', to facilitate their administration of the extensive territories. Boddy's chapter in this volume describes the way in which indirect rule ushered in a period in which there was a great attempt to identify and reinvigorate the different 'tribes' and their 'native authorities'. 'Modern education' was seen as a dangerous force that removed people from their cohesive tribal communities. Yet some of the paradoxes of the British rule are evident here as the British simultaneously encouraged missionaries to go the south and to engage in educational activities. At this time the British also encouraged West Africans to settle, partly to fill a labour gap that had been left by the end of slavery. The demand for agricultural labour also multiplied after the introduction of dams across the Blue Nile and the establishment of huge irrigation schemes; the largest among them was the Gezira Scheme.

The period that followed independence in 1956 was characterized by uneasy political coalitions that fell victim to military coups. All the regimes, such as that following the 1958 bloodless coup, led by Major General Ibrahim Abboud, were dominated by northerners. In 1963, a southern guerrilla movement emerged called Anyanya, which became a rallying point for southern frustrations, and had numerous highly visible and symbolic military successes (see Johnson and Hutchinson, this volume).

The success of the Anyanya led, by some direct and some indirect means, to the fall of the Abboud regime in 1964. The transitional government that followed attempted to include many of the different political interest groups, from members of the Communist Party to the Muslim Brotherhood (led by Dr Hasan al Turabi). Despite some optimism at the inclusion of southerners in government and a period of relative free speech and political organization, the transitional government also suffered from political rivalries, from the continued military successes of the Anyanya and from economic problems.

In 1969, Jaafar Nimeiri came to power in a bloodless coup. Over the next sixteen years Nimeiri aligned himself with factions from many different sides of the political spectrum. Starting out as a socialist, he later formed a close alliance with Hasan al Turabi, and, in 1983, Nimeiri announced that principles of shariah law would be enshrined in the penal code: theft, adultery and murder would be judged according to the Koran; alcohol consumption and gambling were prohibited. This 'Islamization' of the state has been a constant cause of tension and conflict with the more Christian south. At the same time, Nimeiri was becoming closer to the US, which saw Sudan as an important ally in the cold war. In the 1980s, Sudan became the largest recipient of foreign aid in sub-Saharan Africa, and de Waal (1997) describes how the US went to great lengths to ensure that the aid was continued, despite Sudan's constant defaulting on commitments and evidence of economic mismanagement.

In the early part of Nimeiri's rule, the fighting with the south continued and the Anyanya gained in military and political strength. In 1972, a conference in Addis Ababa succeeded in brokering peace and the three southern provinces became part of a self-governing region. The peace did not last long, however, as there was resentment at perceived Dinka dominance in the southern government; ex-Anyanya troops incorporated into the armed forces also caused problems. A new group of fighters emerged in the south, known as the Anyanya II, followed by the formation of the Sudanese People's Liberation Movement (SPLM), whose armed wing, the Sudanese People's Liberation Army (SPLA), was headed by John Garang. The SPLA troops soon took over large sections of territory in the south.

Nimeiri was overthrown in 1985. Following a short period under a transitional government, a coalition government was formed with Sadiq al Siddiq (also known as

Sadiq al-Mahdi) as Prime Minister. The new government began to arm tribal militias to counter the southern rebels, a military strategy that sadly was to become more familiar in the regimes that followed. Tribal militias in western Sudan, armed by the government, made devastating attacks on the peoples of Bahr el Ghazal and Equatoria (Holt and Daly 2000). This period was also characterized by severe famine; food failed to get to the needy because the grass-roots administrative structures had been undermined; the international community failed to respond adequately to calls for assistance; and food and famine were used as a weapon by both the SPLA and the government (de Waal 1997).

A coup in 1989 brought ʿUmar Hasan Ahmad Al Bashir to power, and began a regime supported by the National Islamic Front (NIF) and the Muslim Brotherhood. The regime has been accused of unprecedented repression of its people and exercise of state power. Up until the mid-1990s, the SPLA gained territory and strength, but then certain developments occurred that weakened its position: in 1991, the Derg government in Ethiopia, led by Mengistu Haile Mariam, was overthrown, and the SPLA lost one of its main sources of support and arms. The new government expelled SPLA forces and other Sudanese living in Ethiopia, particularly around Gambela. In August 1991, three SPLA leaders, Riek Machar, Gordon Kong and Lam Akol Ajawin, attempted to overthrow John Garang, and, after failing, became leaders of a splinter faction known as SPLA-United. The SPLA-United faction is sometimes referred to as the Nassir faction, named after the place where Machar, Kong and Ajawin were based. Johnson and Hutchinson's chapters in this volume explore the nature of the conflict that followed between these groups of southerners.

Conflict continued throughout the 1990s, and the Bashir government has retained power, despite many problems. When the Bashir government supported Iraq in its invasion of Kuwait, it lost one of its most valuable economic resources, the remittances from Sudanese workers in the Gulf states. To the west, Darfurian rebels have been active since the early 2000s. The government has been accused of supporting another tribal militia, the *janjawiid*, and encouraging them to rout the rebels. As in other conflicts, there has not been any distinction made between rebel fighters and civilians, however, and the fighting has been characterized by extreme violence and brutality and by multiple atrocities. Darfur residents have fled to refugee camps and across the border into Chad, where their presence has caused political instability. The African Union has sent troops to act as peacekeepers, but their numbers and resources are inadequate. At the time of writing, the United Nations continues to hold discussions on intervention to halt what the US government has described as 'genocide' and the UN is speaking of 'the greatest humanitarian disaster today'.

Although there is no contribution by a specialist on Darfur in this collection, it is worth pausing for a moment to examine some of the literature and to see how this recent conflict is connected to other conflicts, including the conflicts described in this collection, and to discuss whether any common patterns can be identified. The examination presented is based on studies by Kurt Beck and Alex de Waal (Beck 2004; de Waal 2004, 2005) and, for the earlier history, on the almost encyclopaedic volume by Ulrich Braukämper (1992).

In the popular media the conflict in Darfur is described as one between 'Muslim Arabs' and 'black Africans'. The implication is that the black Africans are much more recently and less profoundly Muslim, if at all. Since political decision-makers draw most of their knowledge from these popular media (and then re-feed this knowledge to the media with the effect of a positive feedback), this dichotomy, whether wrong or right, has developed into a political truth in the minds of many of the relevant actors.

Examining the history of the religious and racial categories implicated in the dichotomy, it becomes clear that it is far too simplistic. Islam spread from northern Africa southwards much earlier in Western Africa than it did in the east of the continent. Sub-Saharan Africa was largely Islamic in its Western reaches at a time when Islam was still far from prevalent further east. For example, in West Africa, Takrur (in present Senegambia, eleventh century) and the empires that had their centres in the bend of the Niger (modern Mali), namely Ghana (ninth and tenth centuries) and Songhai (fifteenth to sixteenth century), were early Islamic empires that had accumulated legendary wealth and reached a high level of Islamic scholarship. At the same time, regions on the same latitude in the Nile-Sudan were still Christian, such as the Nubian kingdoms and, further east, the heartland of Ethiopia, which is still basically Christian today. Vast areas in between were characterized by African beliefs and rituals. The old routes of the haj avoided eastern Sudan. They passed through the Sahara to the Fezzan (in modern Libya) and then along the Mediterranean coast and the Red Sea to Mecca. Gradually, Islamization expanded along the Islamic belt in all directions, but especially from west to east. The assumption is often that this process must have taken effect in the opposite direction due to the greater proximity of Eastern Africa to the Arabian Peninsula, but this assumption is not supported by history. Over the course of the shifting of the haj route to the south – over Sennar and Sinja to the Red Sea – Darfur became an important stage in the pilgrimage route. There is no reason, therefore, to assume that Islam in Darfur is in any way inferior in age or importance to that of the Nile-Sudan, or to attribute it to the Arab factor. Certain Arab tribes, centred in Kordofan, such as the Kababish (Asad 1970: 13) and the Kawahla (Beck 1988: 31), both Baggara or cattle-herding Arabs, have expanded recently to the west. How strong their faith is and what role Islam plays in their tribal policies (apart from legitimizing certain arguments and alliances) are questionable. Other Baggara Arabs, like the Humr, appear to have moved westwards from areas now located in Chad. Their movements were motivated by the desire to avoid demanding sultans and to move away from their centres of power or into the domains of less demanding rulers (Cunnison 1966: 3f.). However, in the kingdoms of the Sudanic belt, comprising Darfur, there is no evidence that the Arabs were more Islamized than the non-Arabs. In fact the Arabs were typically under the rule of more educated political elites with a higher standard of Islamic learning who were non-Arabs. At no point in recent or earlier history could it be said that the front line in Darfur is between Muslim Arabs and (implicitly non-Muslim or newly Muslim) black Africans. Rather than Darfur being on the receiving end of the expansion of 'Arabic-Islamic civilization' coming from the Nile, it is the Nile-Sudan that owes many elements of its Islamic tradition to 'Africans' from the west and to Darfur's influence.

The popular dichotomy also implies a racial component. The 'Arabs' are imagined to have a lighter skin than the 'black Africans'. But on a closer look this does not hold true either. In the Nile valley one finds 'Arabs' of many different shades of pigmentation, including people who are much darker than most 'Africans'. Among the Baggara of Kordofan and the Arab nomads of Darfur, however, the lighter shades of pigmentation can be found much more rarely. This racializing dichotomy therefore has no correlate at all in the observable reality. The Arab militias (*janjawiid*) and the many oppositional militias cannot be distinguished on grounds of pigmentation. There may have been Arabs who at the time of their arrival in Kordofan and Darfur had lighter skins compared with the speakers of Nilo-Saharan and Niger-Kordofanian languages in the region, but these tend to have lost that particularity through intermarriage with Fulɓe[1] and concubinage with Nilotic slaves. Furthermore, in many regions in Africa not only one's biological father is referred to and addressed as 'father' but also the shaikh who led a person to Islam and gave him or her an Islamic name. It would be hard to ascertain how many of the patrilinear genealogies going back to the Quraysh, the tribe of the Prophet (curiously, other Arabic tribes are rarely mentioned in the genealogies), are 'Arabic' only by the mere spiritual connection. The Prophet himself, by the way, was not too fond of genealogies and often stressed the equality of all Muslims.

Above, we have already briefly discussed the Mahdiyya from a Nile valley perspective as a revolt against Egyptian rule. Much of it, however, was a West African and western Sudanese phenomenon. Expectations of the Mahdi, the 'rightly guided' leader of the Muslims, spread from West Africa to Sudan in the late nineteenth century. A Mahdi was found and he succeeded in conquering Sudan (1874). But shortly after (1876) he died, and a caliph (*khalifa*) was named to replace him, following the example of the life of the Prophet. This *khalifa* was a West African, like many of his followers, who came from West Africa and Sudan, including Darfur and Kordofan. After the recapture of Sudan, which ended with the Battle of Omdurman (1898), the sons of the high-ranking Mahdists quickly became the new educated elite. The new college was named after the Mahdiyya's most prominent victim (Gordon Memorial College, today the University of Khartoum). Until today, the legacy of the Mahdiyya influences one of the two or three most important political currents in Sudan.

Independent rule was maintained in Darfur longer than in any other part of present-day Sudan. Darfur was conquered only in 1923. The fronts in Darfur's anti-colonial fight were drawn – needless to say – not between 'Africans' and 'Arab Muslims' but between the defenders of Islam and the 'infidels' (the British).

A renewal of western Sudanese elements in the political Islam of the country followed the dramatic change in 1989, when the National Islamic Front (NIF) came

---

1. According to Braukämper, this union also led to the cow-herding culture of the Baggara 'Arabs'.

to power. Its chairman, Hassan al Turabi, had his power base in the west. He served as a figure of identification for Islamists across the whole world. He had the ominous reputation of being the 'Black Pope', and, since Osama bin Laden had been a guest in Sudan until 1996, Turabi served as the embodiment of the enemy to Americans. Since he was placed under house arrest in 2000, he is said to have ties to the JEM (Justice and Equality Movement), a major rebel group in Darfur, and that many of his partisans have formed the JEM is undisputed.

By now it should be evident that the aforementioned, widely popular dichotomy of 'black Africans' versus 'Muslim Arabs' does not provide categories adequate for describing the social reality or the reasons for conflict. But, if it is erroneous, how did it originate and why does it prevail?

Possible explanations for its origins can be found on a global scale and on a more local one. Globally, the idea of dangerous Muslims threatening non-Muslims has long been used to explain behaviour and outcomes. It can be traced to the early modern period in Europe and beyond that to the Middle Ages. The Reconquista of Spain, fuelled by the ideas of the crusades from many centuries before, directly precedes the expansion of European powers onto other continents. Africa was explored starting from the eastern and southern coasts inwards in competition with Islamic states and trade networks that had formerly sealed the continent off from Europe in the north. On a local level, colonialists, especially the British, often cooperated with Islamic potentates, celebrated Islamic holidays and perceived Christian missionaries as bothersome. And yet the reality is that, throughout the colonial period and after, the bulk of the taxpayers and churchgoers in Europe viewed Islam and the Christian and post-Christian 'civilizations' to be competing models. Since 11 September, the world is now completely caught up in these dangerous and simplistic dichotomies. They are so popular that their mere appearance anywhere should be reason enough to question them.

The other origin is in Sudan itself. For some actors in the conflict in southern Sudan (Johnson, this volume) the strategy of dichotomizing 'black Africans' and 'Arab Muslims' proved quite successful. In southern Sudan, the notion that English-speaking Christians of African descent are oppressed, exploited or enslaved by Arab Muslims can be a useful fiction for 'crusaders'. That is the term the regime in Khartoum uses for foreigners, mostly representatives of American Protestant churches who have allegedly financed and fuelled the conflict. Similarities between the ideas of the religious right in America and the crusaders of the tenth and eleventh century are indeed hard to refute. The chapters in this collection show that there is no neat division between 'Muslim north' and 'black Christian south'. The command language of the SPLA is Arabic. During the last years, the fighting in the area was dominated as much by conflicts between different factions of the SPLA, led by Nuer and Dinka, as it was by conflict with the north. The government in Khartoum was only one of the parties forging ever-changing alliances and conflict. Since oil production started in Sudan several years ago, the desire to gain power over the oilfields provides a far better explanation for the conflict than theological position (Christian vs. Muslim).

How then can we explain the success this dichotomy has had in spreading and gaining wider currency? It is a strange success story as it has been found inadequate for the case to which it has been recently extended (Darfur) and only partly adequate for describing the case from which it was derived (north-south conflict). Under the heading 'a size theory of identification', Schlee (2004a: 136; see also the introduction to Volume I), on the basis of examples other than Darfur, has proposed that it is necessary to pay attention to identity discourses with regard to the sizes of the groups and alliances these discourses propagate. In other words, it is necessary to look at identification strategies from the perspective of the group sizes that result from them. There are conditions under which it seems rational to make the group limits more stringent (arguing for exclusion of group members who are not needed for maintaining the resource base of the group, and with whom one therefore does not want to share these resources). Other conditions will promote expanding the 'we'-group and alliances (arguing for inclusion in the face of a threat or when trying to gain access to resources). So it makes sense to analyse whether the size and power[2] of the group or the alliance expected to be reached after successful implementation of an identification strategy played a role in choosing this particular identification strategy. This analysis should be carried out in comparison with other factors.[3]

By analysing identification strategies in light of the resulting alliances in Darfur, while at the same time stepping back to view the broader relations, an answer is found to the question of why the dichotomy between 'Africans' and 'Arabic Muslims' is so successful. Experts on the region have also come to a similar answer without having to learn Schlee's jargon first. Alex de Waal points out that it makes sense for one party to propagate its 'Arabic' identity and still, despite the atrocities committed,[4] claim to be Muslims, because this pan-Arabic and Islamic identification elicits the help of Libyan president Muammar al-Gaddafi. On the other side, the argument that American support can be mobilized by people who speak just as good Arabic and are also Muslims, through the argument that they are 'African' victims of 'Arabic-Islamic' persecution, requires a bit more explanation. Alex de Waal distinguishes between four types of Americans:

1. The US government.
2. The religious right (only partly identical to 1).
3. Liberals and human rights activists.
4. Americans with African ancestors.

2. The power is dependent not only on the demographic size but also on technological factors.
3. Those factors might include historical plausibility, which is flexible but of course not unlimited; the virtuosity of the leaders (their knowledge and social skills); or psychological factors such as collective trauma. Countless historical examples prove, however, that these factors do not often seem to hinder the assimilation to or forging of alliances with former enemies.
4. Atrocities are often committed in response to previous atrocities, and lead to further atrocities. In situations like these, reassigning the blame is not the trick, but stopping the cycle of violence.

In the Darfur conflict, the US government can be assumed to be a single coherent actor, at least with respect to its foreign policy towards Sudan.[5] Throughout the 1990s the US deemed the regime in Khartoum a 'rogue state' and aimed to overthrow it. Advances by the Sudanese government to help in the fight against Al-Qaeda were rejected on the grounds that the regime in Khartoum itself was terroristic. A more prudent policy might have helped to prevent the 11 September attacks. In November 1997, the US imposed a complete trade embargo against Sudan, excluding the export of gum arabic, an ingredient of Coca-Cola. Later *Der Spiegel* reported that Osama bin Laden had brought the gum arabic export under his control prior to his expulsion from the country in 1996. Believing they were fighting international terrorism, the US boycotted the entire Sudan except for the business interests of Osama bin Laden. In 1998 the Americans bombed a factory producing malaria pills in Khartoum. As a justification they claimed that toxic gas had been produced there, or that it had belonged to Osama bin Laden. Sudanese friends assured Schlee that they could name ten targets belonging to Osama bin Laden, but that this particular factory was not one of them. The Carter Foundation later concluded that it was a falsely identified target. An admission of guilt or even payment of damages by the US has never come.

This hostile but incompetent policy was later revised when the peace talks on Sudan taking place in Kenya[6] started to move in a promising direction. There is now a peace treaty (9 January 2005, Nairobi, called the Comprehensive Peace Agreement (CPA)), which includes a six-year term of shared power between the former conflict parties (power sharing), the division of oil interests (wealth sharing) and the holding of a referendum after the end of this term to decide on a unified state or a peaceful separation.

Peace in Sudan, however, is not necessarily in the interests of the 'crusaders' within the religious right in the US. It is not surprising that the 'crusaders' are supporting the opposition in Darfur, who, in turn, facilitate this alliance by exaggerating their 'African' characteristics and correspondingly blanking out their Islamic identity. Other forces, usually not in the company of the religious right, are also part of this alliance. Liberals and human rights activists have – in part rightly so – taken up a position against the Sudanese government for a long time, and are thus tempted to assume that enemies of this government are better or different. African Americans also rather tend to identify with the 'Africans' than the Muslim 'slave hunters'.

These alliances also have to be considered against the regional backdrop, including Chad.[7] Cross-border ethnic groups, including the Arabs but also a vast number of

---

5. The Sudanese government cannot be perceived as a monolithic entity. Parts of it (like parts of the northern opposition) have supported the *janjawiid*, but it is doubtful that pressure on them to disarm the *janjawiid* can achieve much. Even the elements willing to do so might not be capable of carrying this out. Among the numerous factions of the *janjawiid* there are many of which not even the local leaders are in effective control.

6. First in Machakos, later in Naivasha.

7. We are grateful to Andrea Behrends, researcher at the Max Planck Institute for Social Anthropology, for sharing her knowledge on Chad.

other groups, such as the Tama, the Masalit and the Zaghawa (of the last of which Chad's president is a member), play a role in this context, as do some aid organizations with operative bases in Chad. The situation is further complicated by conflicting American and French oil interests in Chad, Cameroon, Nigeria and other regions, as well as by the problem of oil transportation routes. Political weaknesses are exploited. The Chadian government only obtains 10 per cent instead of the usual 50 per cent of the oil profits from the licensees. Perhaps this is part of the explanation for the instability in the region. There are strong interests that do not foster self-reliant governments or peaceful and prosperous states. Much of what goes on may never be publicly known. But among the few certainties we can have is that the ethnic-religious macro-categories of 'Africans' versus 'Arabic Muslims' are historically inadequate. They were not pre-existent; they do not date back to a time before the conflict, and thus they cannot be used to explain the cause of the conflict. Their origin and their use are as labels for identification that have emerged in the course of the conflict.

The fact that they are favoured by journalists can easily be explained by the thirty- or ninety-second slots the TV journalists are allocated for their stories. They are also, in general, quite laborious to differentiate in detail. An account like the one at hand here, however short and only comprising a summary of already existing secondary literature, would still be too long and complicated for the daily press. In this situation the circulation conditions for simple, albeit wrong dichotomies are ideal. They capture the imagination of many, among them political decision-makers.

On borders

The histories of particular nations are not necessarily very instructive regarding what happens at the margins of these national units, where territories have been contested and boundaries drawn and redrawn more to serve the interests of bureaucrats at the centre than the interests of the people who live there. In this corner of North-East Africa, the boundaries between Kenya, Uganda, Sudan and Ethiopia have shifted, depending on the willingness of particular colonial governments to take responsibility for administering these areas and depending on the relations and competition between the colonial state governments. These areas were seen in the main as logistically difficult and expensive to manage. It is notable, however, that, since the end of colonialism, the boundaries have remained remarkably stable (Clapham 1996).

During colonial times, Ugandan territory was much larger than it is today. The boundary between Kenya and Uganda was moved eastwards in 1902. Prior to that, the line had been drawn boldly through the middle of Lake Turkana (then Lake Rudolf), undulating south through much not yet effectively controlled land to what is now the boundary with Tanzania (then Tanganyika or German East Africa). This line included Lakes Naivasha and Baringo on the Ugandan side. The area south of the River Turkwell (passing Lodwar in Turkanaland and flowing into Lake Rudolf) was transferred to Kenya in 1902, but the area north of it, to the present boundary in the west, was transferred only in 1919 (Barber 1968).

A part of southern Sudan, the Ilemi Triangle, has been under more or less continuous Kenyan administration since colonial times (Tornay 2001: 39).

'Continuous' in these peripheral areas has not meant any consistent degree of state penetration or efficacy. But, whatever state presence there was, it was that of Kenya and not that of Sudan. What Barber says of Karamoja in north-west Uganda may be true for all these border areas, which have been claimed by different powers at different times, sometimes with hesitation and often with little success, and often only in order to stake a claim: colonial incorporation was 'an attempt to acquire authority on the cheap, to state rights without accepting obligations' (Barber 1968: 11). The impact of the colonial borders and administrative institutions are only some of the, albeit very important, factors that have shaped the nature of identifications and alliances in the region.

Uganda

There is only one chapter in this volume that focuses specifically on Uganda, that by Gray, which needs to be placed within the context of the political history of Uganda. Uganda's political history is of wider significance to this volume, however, as the relations between Uganda and Sudan have influenced the nature of conflicts, identifications and possibilities for peace on both sides of the Uganda-Sudan border.

At the turn of twenty-first century, areas of Western Equatoria (Sudan) appear to be more orientated to the Ugandan state than to the Sudanese. In Yambio and Nzara towns, for example, in 2007, the Ugandan shilling was still the currency, not the Sudanese dinar or the new Sudanese pound. Individuals travelled to Ugandan towns such as Arua to buy basic provisions (such as hoes), if they wished to be tested for HIV/AIDS or to obtain other forms of health care. Seen from the Sudanese side of the border, therefore, the Ugandan state appeared to be functioning well, but this appearance disguises the very troubled and turbulent history of Uganda and the levels of violence experienced by its citizens.

From 1894, Uganda was a British Protectorate, and, from the beginning, the state was characterized by an imbalance in investment, development and benefits enjoyed by people in the north and the south, and also between different groups within the south. In the south, when the British arrived they found the powerful kingdom of Buganda, with its own king, the Kabaka, and his council of 'ministers', the Lukiiko. The kingdom was already relatively well armed and educated (Gertzel 1976). The British entered into an agreement with the new leadership of Buganda, which was given a certain degree of autonomy under an indirect rule arrangement. The British also recognized the other southern kingdoms, those of the Toro and Ankole, but did not give them the same privileges. Finally, the British were also much more aggressive towards the historic rival of Buganda, the Bunyoro kingdom; the Baganda assisted the British in controlling Bunyoro and gained position and territory in return. As a consequence of their original advantage and favoured position, much of the early development and industrial activity of Uganda took place in Buganda, where the towns of Kampala and Entebbe are now situated.

One axis of inequality between these groups was institutionalized by the colonial state, therefore. A second main axis of uneven development in the country corresponded to the line between north and south usually drawn along the River Nile, from Lake Kyoga to Lake Albert. To the south and south-west of this line, where the kingdoms are located, the people are described as Bantu; to the north and north-east they are described as Nilotic, Nilo-Hamitic or Sudanic. Consistent with the colonial policy of divide and rule, the northern areas, where the Langi, Acholi and West Nile people are situated, were mainly considered as reserves of migrant labourers for the central-south, which was promoted as a cash-cropping and industrial area. Additionally, while many civil servants came from the south, the north became the main source of recruits for the army, police and prisons. At independence in 1962, therefore, the country was structured into areas of unequal economic development, which were regional, but also to some extent related to and perceived as being connected to cultural and ethnic identity (Gertzel 1976). More than thirty thousand Asians were also engaged in commerce and industry in the south. In addition, there had been a long history of Christian missionary activity in the country, with French missionaries in the nineteenth century encouraging conversion to Roman Catholicism, while British missionaries encouraged conversion to Protestantism (Karugire 1980). Religious identity became another main axis of difference, and, following the British promotion of Protestantism during the colonization process, Protestantism became the religion of the establishment. Islam, which had come earlier through traders from the coast or the north, was also present in the country, but as a minority religion. From the beginning, religious practice and identity have often cross-cut as much as they have coincided with ethnic identities, but they have been a further factor defining the lines between conflicting parties in regional, national and international political struggles.

Accounts of Uganda's history include those of Gertzel (1976), Mamdani (1976), Karugire (1980), Mutibwa (1992), Behrend (1999), Reno (2002), Knighton (2003) and Rake with Jennings (2003).[8] These accounts describe a history of post-colonial Uganda as 'one of violence and counterviolence' (Behrend 1999: 23). Since independence, the majority of regimes have come to power through the use of military force and have continued to be closely allied to the army or to unite the roles of head of state and head of army. The history is one of power struggles and shifting alliances in which the use of violence, often by the state, has been institutionalized.

At independence, the first Ugandan government was made up of a coalition between the Kabaka Yekka (KY – 'Kabaka only') party and the Ugandan Peoples Congress (UPC). The Buganda-based KY was led by the Kabaka, Edward Mutesa II and the UPC was led by Apollo Milton Obote, a Langi from the north. Soon after independence cracks began to open in the uneasy north-south partnership. Obote consolidated his power through judicial and administrative reforms, politically motivated appointments and attacks on trade unions and other civil society organizations. In 1964, the KY-UPC alliance was formally dissolved and Obote dismissed KY members from government positions.

In 1965, dismissed and alienated KY and UPC members accused Obote, together with Colonel Idi Amin, the Deputy Army Commander, of embezzling

---

8. We are grateful for the comments on Ugandan history from Godfrey Asiimwe, Joe Powell and Glen Rangwala. The authors remain entirely responsible for the content of this introduction, however.

money, gold and ivory from Congo. In response Obote ordered an attack on the Kabaka palace. Led by Idi Amin, the attack, known as the battle of Mengo, has been described as the 'first major bloodbath of postcolonial Uganda' (Mutibwa 1992: 39) and was followed by more harassment and killing of Buganda people.

What started as harassment of the Baganda people was soon extended to anyone who was considered to harbour resistance to Obote's regime. In 1967, the attack on the power of the southern kingdoms was completed, as a new constitution abolished all forms of kingship and declared Uganda a republic. The republic was a repressive one, however: no opposition parties were permitted, and violence, torture and detention were commonplace. Economic problems were also rife.

In the last years of Obote's rule, Obote's and Amin's sources of military strength became increasingly divided along ethnic lines. By 1969, according to Mutibwa (1992), Obote drew support from the Acholi and Lango sections; Amin, himself from West Nile, drew support from the West Nile peoples, mainly from the Kakwa, Lugbara and Madi (Allen 1994). In 1971, Obote accused Amin of corruption, in a move that appeared to be the first step of removing him from office. In response, when Obote went to a meeting of Commonwealth leaders in Singapore in January 1971, Amin seized power.

Amin's rise to power was popular initially. He was heralded as a liberator and a peacemaker. But the violence continued: thousands of Langi and Acholi soldiers, seen as Obote supporters, were massacred in their barracks. Amin's term of rule was initiated therefore with the calculated killing of a large group of people he viewed as a threat. One year into his office he also expelled all the Asians from the country, on the grounds of promoting the economic interests of the Ugandan people. No positive economic results were felt, however, and the heavy-handed manner in which the Asians were expelled, in which they lost much property and were abused and displaced, has been heavily criticized.

The appalling story of the Amin years is well known and impossible to do justice to here. Outwardly charismatic and successful on the international stage (becoming Chair of the Organization of African Unity for example), his regime at home was characterized by the brutal murder of prominent people and ordinary civilians, by torture, detentions and dismemberment of individuals and by the general destruction of lives and livelihoods, often at the personal order of the President. This rule went on for eight years, until, finally, when Amin invaded Tanzania, Tanzanian troops counter-attacked. With the help of rebel forces known as the Ugandan National Liberation Army (UNLA), Amin was defeated in 1979. With the fall of Amin, the Acholi and Langi sections within the UNLA carried out a massive attack on West Nile, destroying the town of Arua and killing citizens and soldiers, in revenge for the West Nile attack on Acholi and Lango almost a decade before.

In 1980, a general election was held, and, despite the controversial results, Milton Obote and the UPC returned to power. This prompted the outbreak of a guerrilla war. By 1985, the Yoweri Museveni-led National Resistance Army (NRA) had weakened government forces significantly and Obote was overthrown in a coup. In 1986, after an interim period under the rule of General Tito Okello, an Acholi, the NRA took Kampala, bringing Yoweri Museveni to power.

It is clear that the successive regimes in Uganda have been highly militarized and associated with one faction, north or south, or ethnicity. Massacres have led to counter-massacres. The associated experiences of conflict and violence have been profoundly disruptive for individuals and societies. When Yoweri Museveni came to power, for example, thousands of demobilized Acholi soldiers were sent back to their northern territories. Behrend (1999) describes how these soldiers experienced economic, health, psychological and social problems on return, and they often vented their frustrations on their home communities. Returnee soldiers and others found some comfort in new hybrid religious movements, in this case the Holy Spirit Movement, led by Alice Lakwena. Through the movement, problems were understood as punishments by spirits for the violence that soldiers had committed. Through the movement's rituals, the soldiers could be cleansed from the blood they had spilt, thereby avoiding the revenge of malevolent spirits. The blessings and keeping to a strict behavioural code were also believed to give members of the Holy Spirit Movement invulnerability in their violent conflict with the NRA. Accounts of their battles told of how they simply walked into the fire of their enemies; anyone who fell was thought not to have kept the code correctly (van Acker 2004).

In 1987, the Holy Spirit Movement was defeated, but it was later reborn as the Lord's Resistance Army (LRA), under the leadership of Joseph Kony. The LRA has terrorized the north with its atrocities, including mutilations and killings, and the abduction and forced conscription of children into its army as fighters, 'sex slaves' or porters (van Acker 2004). Human Rights Watch (2002: 5) estimated that, overall, the LRA 'has abducted an estimated ten to twelve thousand children under age eighteen, one third of the total abductions, since it began its mobilization. Of these, one third were under age twelve when abducted.' The people involved in the LRA are mostly Acholi, and Acholi have also been the main victims of their violence.

In order to protect the local population, the Uganda People's Defence Force (UPDF – the current government military force) ordered people into displaced persons' camps or towns. By 2002, over half a million people were estimated to be living in displaced persons' camps, where, unable to cultivate land or pursue other livelihood activities, they have become dependent on relief. The displaced persons' camps do not appear to have provided the safety that was intended, however. The LRA have often raided camps, where they have found a concentration of people whom they have attacked, killed or abducted. They have also found relief supplies there, which they have stolen or burnt. In addition, there have been many accounts of attacks on displaced persons' camps by the UPDF itself (Human Rights Watch 2002); the people's supposed protectors have often proved to be their attackers.

To the east of Acholi are the people of Karamoja, who are described in Gray's chapter in this volume. They have not been among the powerful players in the struggles to control the centre of political power. The people in Karamoja – the Karimojong, Jie and Dodoso – are pastoralists and agro-pastoralists. Although they have not been heavily involved in the conflicts described above, they have a long history of violent raiding against their neighbours in Uganda (the Langi, Acholi, Teso and Sebei), against their neighbours in Kenya (Turkana and Pokot), between themselves and within different territorial sections of Karimojong, Jie or Dodoso.

Overall, this conflict has also led to high levels of mortality and the displacement of hundreds of thousands of people (Knighton 2003). In the past, according to Knighton, the government was content to ignore this conflict, as it was mainly directed against groups such as Langi and Acholi, which at times had represented a political threat. The raiding kept their potential political opponents busy and weakened them. But, since 11 September 2001, any area of insecurity outside the control of the state is viewed as a potential source of or refuge for terrorists, and new funds have been made available to deal with this conflict situation. The result has been a large-scale disarmament programme, led by the UPDF.

As in Acholi, the UPDF activities in Karamoja have been characterized by human rights violations and atrocities visited upon the local population. UPDF burning and shelling of houses, raping of women and forced exposure to HIV/AIDS, and beating and killing of old people, women, children and pregnant women have all been recorded (Knighton 2003; Gray, this volume). The perception locally is that the state itself is another 'raider' (Knighton 2003; Gray, this volume), that is pursuing disarmament as a strategy of individual gain rather than as a strategy for peace. The disarmament has left Karimojong pastoralists unable to defend themselves against attacks from the UPDF or from the Pokot and Turkana. At the same time, the Karimojong raids have also become more brutal, as raiders have adopted some of the techniques of the UPDF soldiers, including rape. This is not the first time brutal attacks from state soldiers have been visited on local pastoralists in the name of disarmament, however. Knighton (2003) describes the attacks of the British on the Jie, when local people were crucified or nailed through their foreheads to trees when they failed to give up guns. In this part of Uganda, brutality has long been used by the state 'as a quite effective means of imposing their presence' (Knighton 2003: 437).

Reno (2002) argues that it suits Museveni to keep the army busy in the north or in Congo. While the military is engaged in battles elsewhere, it does not represent the same threat that it has proved to be to previous governments that fell victim to military coups. Personnel involved in Congo may also be able to acquire personal wealth through trading in gold, timber, diamonds and niobium (used in mobile phones); and, to a certain degree, trade in Congolese resources may be a useful boost to the beleaguered Ugandan economy (Reno 2002). Thus, Reno suggests that the government does not have the power to do anything about the military's extra-legal activities, and that it may also serve its purpose to turn a blind eye to them.

Despite these problems and the ongoing violence, Museveni is currently seen as the most successful leader in modern Uganda. He has brought peace to the south, if not to the north, and is the 'darling of the donors' (Doornbos 2005: 7). After years of economic problems, a sustained level of 3 per cent growth in the Ugandan economy is being heralded as a result of Museveni's successful negotiation and implementation of debt rescheduling and 'Highly Indebted Poor Country' (HIPC) initiatives, supported by the World Bank, the IMF and the Paris Club (UNDP 2003). In the late 1980s, Uganda was seen as the worst HIV/AIDS-affected country in the world, but it is now one of the few countries in Africa where the pandemic has been controlled and is subsiding. Studies of the success of the Ugandan approach to HIV/AIDS have pointed to the difference that the government has made by its quick response and the way in which it has worked in partnership with civil society organizations and the international community (Allen and Heald 2004).

Little is known about HIV/AIDS rates in the north of the country (Allen and Heald 2004), however, and peace and economic growth appear not to be enjoyed there. Doornbos (2005) warns against seeing Museveni as a politician who has broken the mould of personalized politics in Uganda. The constitution has been changed to allow him to stand for a third term of office. Multiparty politics are now permitted, although the 2006 elections were overshadowed by charges of treason and rape against the main opposition candidate, Dr Kizza Besigye, and the legitimacy of these elections has been contested. Overall, politics and livelihoods are still dominated by the struggles between powerful individuals and between unevenly situated regions and groups; and, directly or indirectly, armed force and brutal conflict appear still to be the dominant means through which these struggles are waged.

Summary of the Chapters

The first of the chapters in the volume are set in southern Sudan and northern Uganda, and explore the changing politics of animosity and alliance between factions of large so-called 'tribes'. They relate to the Nuer and the Dinka, who follow segmentary patterns of identification, and the Karimojong, an age-grade society of the Karamojong group, which is organized along territorial sections. All practise some form of agriculture, but they value cattle-keeping most highly. Raiding and counter-raiding for cattle, as the main form of wealth, has been and remains a norm.

In southern Sudan, the politics of inter- and intra-group raiding and alliance has been exacerbated by the politics of the civil wars that have raged between the north and south of the country. The first chapter, by Johnson, outlines the key events in the first civil war, 1956–72, and the second, 1983–2005. The conflict in the south has frequently been represented and understood as inter-ethnic or intertribal, between the Nuer and the Dinka, and has drawn on theories of segmentary lineage opposition between these two main groups. Johnson argues that this theory detracts from understanding the other causative factors and also results in naturalizing the conflict. It fails to recognize the way in which the different factions in the conflict are made up of people from different 'tribes' and sections of 'tribes'. A better understanding of the reasons for the conflict, he contends, can be found in the *zaraa'ib* system of the nineteenth century, through which the Egyptians expanded their control over these regions by working together with local intermediaries. Individuals enhanced their position and power through cooperation with the Egyptian state, and gained new followers who wished to benefit from their patronage. Such a 'system' benefited the powerful and ambitious, who found opportunities for advancement within it, and gave protection to the weak. Such transformations in political authority, where new elites were formed through their involvement with an external, often colonial, power, are familiar from other parts of Africa (Iliffe 1979). In the south of Sudan, the formation of these elites also cross-cuts the politics of group alliance and opposition based on the descent principles of the segmentary lineage.

Johnson's analysis of the politics of different factions in the civil wars highlights the similarities with the political economy of the *zaraa'ib* system. Different factions form and dissolve for multiple reasons, including the charisma and power of individual leaders and the alliances formed around access to resources. Most recently the struggles between different factions have been shaped by competition for control over oil. They have also had an international as well as a national dimension, and the different factions continue to be ethnically mixed. The analysis demonstrates that local conflicts continue to be enmeshed in the politics of national and global conflicts, and yet it still suits many observers to portray the struggles and conflicts existing in this and other parts of Sudan as 'local', 'intertribal' and 'cultural'. Johnson's work suggests that, where the different factions involved in the conflict do take an ethnic form (and undoubtedly, to some degree, they do), this results from opposing groups becoming involved in national and international conflicts and competition for resources, rather than from pre-existing ethnically based motivations.

The last point is also made forcefully in the second chapter by Hutchinson, who continues the investigation of the conflict in south Sudan. She examines indigenous analyses of the conflict taking place, in which 'sharpening ethnic animosities [were seen] to be the result rather than the cause of the conflict'. Her work focuses on a grass-roots peace-building initiative that attempted to reconcile different warring factions following the split in the Sudanese People's Liberation Army (SPLA) in 1991. The peace-building meeting was organized by a church body, and the idea that the conflict was ethnic, Nuer against Dinka, was taken for granted and institutionalized by the meeting organizers, for example, through the positions in which delegates were invited to sit. Hutchinson uses the text of the debates at the meeting to show the way conflict was viewed by local people in their own words; she allows us to hear the voices that are rarely made audible, those of pastoralists and others caught up in the conflict. The majority of participants rejected the idea that the conflict was ethnic and pointed instead to the roles of leaders, to the struggles for resources and to national government policies and military activities, underlining Johnson's argument that the conflict was part and parcel of wider conflicts and struggles.

Hutchinson's chapter also argues that the nature of the post-1991 conflict was qualitatively different from that of the conflict that had gone before. Previous conflict had been short-lived and subject to local ethical codes. Meaningful peacemaking rituals could be carried out. The post-1991 conflict was more brutal and indiscriminate, and women, young children and old people were often injured or killed in the conflict: 'Everyone recognized that this violence had little to do with the daring, cross-border, cattle raids staged by generations of Nuer and Dinka youths seeking to demonstrate their courage and fighting prowess.' The conflict and the militarization of society that accompanied it had long and profound consequences for structures of authority and peacemaking, resulting in the conflicts becoming further protracted and entrenched. Hutchinson's study of conflict in south Sudan has fed into and been influenced by wider studies of conflict that have viewed contemporary conflict as a kind of 'new war', in form and impact (for example, Kaldor 1999).

The next chapter is concerned with the north of Uganda, where Gray examines the experiences of raiding and counter-raiding between the Bokoro and Matheniko sections of the Karimojong. She places this in historical perspective through a calendar of Karimojong events, and through retelling women's own narratives about their lives.

The accounts from Karimojong are horrifying, in the stories they relate of death, mutilation, displacement and rape. The level of brutality experienced suggests that the Karimojong case fits with the 'new war' hypothesis, in that it represents a fundamentally different and new kind of conflict in terms of the number and scale of atrocities committed. However, Gray questions the extent to which it represents a structural transformation in the nature of conflict, and rejects the possibility that this has resulted from the recent arrival of guns in the region. The Karimojong have had guns for a long time, and Gray argues that the conflict was always vicious: 'evidence of pastoralists' brutality is present in history and prehistory, if only we look more carefully at oral histories and events commemorated in local calendars'. She argues that, from a pastoralists' point of view, the arrival of the AK-47 (Kalashnikov automatic weapon) allowed them to be more effective and powerful raiders, but it has not radically transformed the patterns of their raiding practices. The AK-47 helps the Karimojong in the preservation of a pastoralist way of life by producing a 'warrior-herder' image, which deters competitors for resources from attempting to graze their animals or attacking; it allows the Karimojong to maintain their autonomy, vital to the maintenance of a pastoralist lifestyle, which has so long been the target of the state and other modernizing policies. Gray's work also concurs with Knighton's work on northern Uganda, in suggesting that the massive disarmament programmes have left the Karimojong more vulnerable to attack from other groups, and also from the state, which has proved the most effective raider of all (Knighton 2003).

Gray investigates the material and biological dimensions of conflict by exploring the nutritional, health and demographic effects of continuous armed conflict (see also Gray et al. 2003). Armed raids in the past have been viewed by some writers as 'functioning' to redistribute cattle and people during times of drought, disease and ecological stress (for review, see Hendrickson et al. 1998). In her study of the Matheniko and Bokora conflict, she shows that the stronger, more aggressive and 'more successful' raiders, the Matheniko, exhibited higher levels of mortality and lower levels of fertility than the Bokora. With mortality levels for Matheniko children prior to age fifteen of above 44 per cent for most decades, it is clear that the risks intrinsic to the strategy of raiding are too high to sustain, regardless of whether or not it 'serves a function' of redistributing people and resources.

Together the three chapters in this section raise interesting questions about the causes of conflict. For Gray, the Karimojong conflict is primordial and age-old, and has been a means through which identities are produced and defended (however unsuccessfully). For Johnson and Hutchinson, the protracted conflict among the Nuer and the Dinka is a product of more recent events and processes, particularly their integration into the Sudanese state and the international political and economic system. The contrast in approach can be explained by the way the peoples involved are different, with different political histories and relations to the states within which they are situated. But these cases also suggest that neither the narrative that gives primacy to primordial processes nor the one that gives primacy to instrumental processes is sufficient for explaining the conflict. Instead, in different situations, different aspects of these processes become more or less relevant. It is the combination of political and economic factors and historical and cultural factors (such as experiences of and attitudes to raiding) that may make the primordial more or less relevant at a particular time, as discussed above.

The two chapters in the second section of the volume broaden our geographical horizons to the north and north-east of the flood plains and swamps of southern Sudan. They describe communities that rely on irrigated agriculture, using the floodwaters of the Nile (Boddy) and the Blue Nile (Schlee). The community on the banks of the Nile, some 200 km north of Khartoum, is made up of descendants of Christian and Nubian groups, whose society was once matrilineal, but has now become entirely Islamicized and patrilineal. The Blue Nile region has been the home of many immigrants, among them people from West Africa, some of whom came to this area on their pilgrimages to Mecca or on their way back from there.

The chapters in this section shift the analysis from the politics and nature of armed conflict and raiding to a different arena of politics and alliance construction, that of marriage and kinship. Schlee describes this more domestic arena as 'a central concern of social actors' and 'the essence of politics'. The chapters continue a theme explored by Gray, however, in that they examine the extent to which cultural practices and biological processes are enmeshed and co-constituted. In the past, approaches to societies have been influenced by a conceptual division between 'nature' and 'culture' or 'sex' and 'gender', owing in part to the influence of structuralism and in part to a reaction against biological or environmental determinism. These chapters illustrate that the dichotomy needs to be rethought: there is no firm line between kinship – anthropological structure par excellence – and biological processes and outcomes. The effects of abstract, ideational 'rules' of marriage are also physical, practical and emotional processes, and these in turn, are part of the emergence of new structures.

This focus on kinship and marriage patterns is not so common in contemporary anthropological studies. Many anthropological students find the study of kinship systems dry, over-deterministic and giving no room for understanding individual agency and experience. Kinship studies are often characterized by convoluted and abstract debates about different 'types' of systems. For those outside the discipline, the 'spaghetti-like' diagrams are obscure, sometimes hard to follow, even mystifying. Boddy and Schlee demonstrate that the working patterns and rules represented in the diagrams are vital for the reproduction of everyday life in this region, and they show how they provide a framework that has an impact on individuals' experiences and life chances. The conventions stipulating whom it is possible to marry and whom it is not are deeply felt and shape senses of self and other.

Boddy shows that creativity and agency are possible within these structures (which can be read prospectively as well as retrospectively) to construct certain alliances as preferred and desirable marriages. Moreover, kinship patterns have counter-intuitive impacts on the qualitative nature of relations in the region and on the development of more inclusive or exclusive social relations between groups. Endogamy is often assumed to be associated with a more closed outlook to others, but Boddy shows how it emphasizes sameness with others and hence can be used to strengthen ties. Schlee, in his comparative study of northern Kenya and the Blue Nile region, concurs with Boddy's findings, but also demonstrates that the large-scale kinship types, such as patrilineal or matrilineal, endogamous or exogamous, contain within them multiple forms. Each of these multiple forms has different implications for the nature of the social relations to others that they construct, the degree of 'openness' or 'closedness' to other groups, and each has different implications for individual lives and experiences. These approaches to kinship are not exclusively structuralist or biologically deterministic, but explore the co-constituting nature of structure, biology and social relations.

The chapters in the third group share a focus on what has happened to people originally from West Africa who have settled in Sudan. Migration from West Africa along the route of the haj has been taking place for centuries. Large numbers of Fulfulde- and Hausa-speaking migrants have moved to Sudan in response to the British conquest of what is now northern Nigeria in the early twentieth century and in the following decades. Abu-Manga's chapter reviews a successful group of Fallata who managed to monopolize what he refers to as a major 'channel and symbol' of modernity in the Blue Nile region, lorry ownership and lorry driving. Lorry drivers developed into a particular class of highly respected and admired people, especially prominent in the years 1960–85.

For anyone unfamiliar with the region discussed by these volumes, the description of lorries and lorry drivers may seem bizarre. But, while many of these chapters tell accounts of conflict and agony or of far-reaching marriage networks, it is valuable to be reminded of other processes and performances that structure identifications and alliances. Abu-Manga's chapter captures the way in which even the apparently most mundane and functional can become special and imbued with hope and be celebrated with humour and colour. The emphasis here is on style: on how apparently trivial matters such as the dress and behaviour of lorry drivers were read and how it became a highly meaningful text. The lorry drivers in the Blue Nile region were viewed by local communities as members of an 'ultra-modern profession': it was through their behaviour, styles and actions and through the way in which they were admired, praised and sung about that a particular modernity in the Blue Nile was constructed. Where some of the other chapters explore the cosmopolitanism of groups (in the sense of their willingness to be open to others), Abu-Manga's chapter explores cosmopolitanism in the sense of an openness to new technologies and the world.

In southern Ethiopia, bus companies place their music speakers on the outside of their buses, so that it is the people in the villages they pass through, not necessarily the passengers on the bus, who hear the latest tunes they bring from Addis Ababa. When a bus goes through a village, small children run to the road and jive to the music; the arrival of the bus is an event that will become a memory influencing a growing child's sense of place in the wider world. Addis Ababa becomes a place of innovation, music and excitement. Abu-Manga's chapter captures elements of a similar process; it also describes the significance and impact of encounters that people have with new technologies. In the same way as others in south Sudan (see the first section in this volume) or in the Lower Omo Valley (see Volume I) were able to use guns to gain an advantage over other groups, the Fallata gained new prestige and position through their monopoly over transport. In the same way as guns played a part in constructing new ideas of the self, the lorries played into a particular kind of performative identity construction, particularly where masculinities were concerned. The lorry driver's swagger was something to be admired or emulated.

Dereje and Schlee's chapter explores another group of Fallata, the nomadic Fulɓe or Mbororo, who graze their long-legged zebu cattle ahead of and further than others, often occupying and leaving pastures before Arabs or Nuer or Dinka can use them. Little is known about these people, who use their marginality and speed to their advantage. Dereje and Schlee show the extensive range of their movements from north to south in Sudan and across into Ethiopia. They illustrate the way in which, in recent years, the Mbororo have adapted to changing circumstances. Their preferred avoidance of others has become increasingly difficult because of constraints imposed by armed conflict blocking their migration routes or because of competition for grazing lands. One strategy the Mbororo have used is to mobilize the negative stereotypes that others have of them. By finally living up to a 'warrior-herder' image, they are able to continue their existence on the margins of the state and other societies and are avoided by others. A similar process of accepting and performing what to many is a negative 'warrior-herder' image is also described by Gray for the Karimojong.

The final set of chapters examines in more detail new structures such as state administrative units, borders, churches and refugee camps and the impact they have on forms of identification and alliance. The focus also shifts to Ethiopia and Kenya, although the people discussed either are from Sudan, have been deeply affected by the processes taking place in Sudan or have an identity that has been shaped by the Sudan-Ethiopia border itself. Dereje's chapter examines conflict between the Anywaa and the Nuer in Gambela town, Ethiopia. The politics of relations between these two groups of people, historically present on both sides of the state border, has been strongly influenced by the construction of the Nuer as non-Ethiopian 'foreigners'. The Nuer have suffered but also benefited from this designation, sometimes becoming the target of attack, sometimes gaining access to valuable resources in refugee camps, such as health care and education. Dereje demonstrates that local-level conflicts between Anywaa and Nuer are shaped by the political relations between Ethiopia and Sudan, whose governments have waged proxy wars in the region. But he also shows how the groups themselves draw on a longer history of conflicts between each other to justify their conflict. He shows again that neither the 'primordialist' nor the 'constructivist'/'instrumentalist' approach to understanding shifting identifications is true in itself, but that both have a role to play in explaining the protracted conflict. At different times and in different places, emic interpretations of conflict and alliance building draw on discourses that can be viewed as both primordialist and instrumentalist to justify and understand their actions.

Falge also writes on the Nuer in Gambela, and focuses on both indigenous Ethiopian Nuer and Nuer refugees who have fled the conflict in Sudan. Her subject is Christian conversion, significant for its impact on identification and also because of the number of churches that have multiplied in the region in recent years. Like the lorry driving described by Abu-Manga, Christianity has been both a symbol and a channel of modernity, and it has profoundly transformed the orientation and forms of identification of local people. Falge explores different theories of conversion and teases out the impact of new technologies, theological dogma, individual personalities and the wider political and economic context. One of the most fascinating dimensions of her work is the way in which she shows that Christian churches have divided and multiplied among the Nuer, especially since foreign missionaries became much less active in the area. Falge argues that the patterns of fission and fusion of different churches and congregations follow patterns similar to that which might be predicted by theories of segmentary lineage. There are new forms of identification here, but they are built on the foundations and structures of older ones.

While many chapters in this volume touch on the important role played by refugee camps in terms of politics or of individual experiences, Kurimoto describes what life is like inside one. Kakuma in northern Kenya is one of the largest refugee camps and is home to many Pari, with whom Kurimoto worked in Sudan in the 1970s and 1980s. Kurimoto describes the priorities of the Pari and the way in which they envisage their future. He also examines the impact of living in a refugee camp in close proximity to people of many different nationalities and ethnic groups. Far from making the Pari more cosmopolitan, he explores the way the experience of living in a multicultural setting and in close proximity to other groups appears to have, in some ways, emphasized their parochialism. The explanation in part comes from the way in which residence in the refugee camp is 'strictly organized by ethnicity and nationality', which emphasizes certain formalized identities. In addition, it is possible that, faced with the multicultural setting, communities often prefer to look inwards and become closed rather than open. Such a process has been described more generally as localization, and is something that has often accompanied the unsettling and fluid nature of globalization.

The last chapter in the volume is by James, who examines the impact that cross-border journeys have had on many of the processes discussed in the other chapters. As conflict has raged in the area and as governments have changed, bringing with them radically different policies, the crossing and recrossing of borders in search of refuge or a more favourable environment are far from uncommon. James describes the profound transformations that crossing a border can have for individual subjectivities. In these essays, many of the pastoralist groups discussed in this volume have sought autonomy from the state in order to escape their policies and as a livelihood or survival strategy. The state policies have often had the effect of making identities more rigid and competitive. James emphasizes that the state is not always a negative force in people's lives. Many people are also actively seeking a political space where the state will protect and serve them. In addition, while some chapters show how groups have the capacity to exclude or even harm others whom they define as strangers and foreigners, she emphasizes a theme that also emerges from many of the other chapters: people equally have the capacity to give support and strength to others, adopting and making them their own.

Part I
Raiding, War and Peace, Sudan and Northern Uganda

Chapter 1

The Nuer Civil Wars

Douglas H. Johnson

What has happened in Sudan? No sooner did the civil war in the south appear to be nearing a negotiated peace in 2003 when another civil war broke out in Darfur. Events in Darfur, as reported by human rights groups and the international media, invite comparison with outbreaks of earlier 'ethnic conflict' and 'ethnic cleansing' in the Nuer homeland. For the Nuer, as later in Darfur, there have been reports of militia formation, fragmentation and realignment; of systematic civilian displacement and deaths; and of oil interests as a factor behind the devastation (Franco 1999; Verney 1999; Amnesty International 2000; Harker 2000; Christian Aid 2001; Gagnon and Ryle 2001; International Crisis Group 2002; Rone 2003).

These events are the latest in a series of a decade and a half-long trend towards the apparent disintegration of Sudan into tribal and 'inter-ethnic' conflict, overshadowing the broader national issues that precipitated the main civil war in the south in 1983. From 1983 to 1991 the southern Sudanese-based Sudan People's Liberation Movement/Army (SPLM/SPLA) waged a successful war against a succession of Islamist-oriented national governments in Khartoum, bringing almost all of the rural areas of the southern Sudan under its control and even extending the war outside the south into the Nuba Mountains, Blue Nile and even, briefly, Darfur. In 1991 the SPLA broke apart in a split that was allegedly about the aims of the movement and leadership style, but which rapidly took on the appearance of an intertribal war, with the massacre of Dinka civilians around Bor by a Nuer army in 1991, followed by a series of raids and full-scale battles along the Nuer-Dinka borders in Upper Nile and Bahr al-Ghazal for most of the next ten years. The 'ethnic' nature of the conflict seemed to intensify after the formal alliance between the Khartoum government and various dissident southern leaders in 1996/97. The Dinka commander, Kerubino Kuanyin Bol, operated with a supposedly Dinka army out of government bases in Gogrial and Mankien to undermine Dinka support for the SPLA in northern Bahr al-Ghazal, and the government further supplied Nuer commanders, who reportedly raised militias from among their own tribes or sections. It was these militias who fought each other in the Nuer homeland.

These events appeared to justify what governments in Khartoum had been claiming about the 'tribal' nature of the war from the start. 'Ethnic mobilization' has been assumed by many outside observers to be the engine of warfare on the ground in Sudan, whatever the wider political issues. Before 1991 they referred to 'the Dinka-dominated SPLA' almost as much as they did to 'the Christian and animist south'. The split in the SPLA precipitated by the breakaway of the Nasir commanders only seemed to confirm this ethnic explanation, and it has been common to present that split in terms of a 'Dinka' SPLA being opposed by a 'Nuer' SPLA-United (or Southern Sudan Independence Army (SSIM), South Sudan Defence Force (SSDF), Sudan People's Defence Force (SPDF), or United Democratic Salvation Front (UDSF), as it has also styled itself). The proliferation of Nuer militias in the Sudan oilfields area only carried on the tendency.

The structural opposition of Nuer to Dinka and the 'predatory expansion' of segmentary lineage systems are entrenched features in the secondary anthropological literature on the peoples of southern Sudan, and it is natural that anthropological commentary on the war would draw explicitly or implicitly on these ideas. Some anthropologists see recent events as evidence of the inherent weakness of segmentary lineages, where those who embrace 'an ever narrower and shorter-term perception of self-interest' are 'divided, pitted against [themselves], and picked off bit by bit by an aggressive and persistent foe' (Cole and Huntington 1997: 79–81). A more nuanced interpretation places the same events in the context of a global trend in the militarization of ethnicity (Hutchinson 2000).

In all the reporting about the SPLA split since 1991 and the subsequent disintegration of the breakaway faction, there has been very little stress placed on the fact that the division within the SPLA generated, almost from the very start, a Nuer civil war. It is my contention that attempts to rally the Nuer politically and militarily have failed, because the 1991 split precipitated a series of interlocking civil wars among the Nuer themselves. The attempt to mobilize Nuer in the escalating fighting along the Nuer-Dinka civilian fault line only widened fissures within Nuer society. Tensions that might at other times have been described as 'tribal', in being confined to a specific locale and regulated or resolved through the structures of tribal administration, have been exacerbated by the national civil war and cannot be explained or analysed in exclusively tribal or 'ethnic' terms.

In this chapter I want first to describe the progress of the Nuer civil wars, focusing particularly on how the superimposition of national conflict has affected local confrontations, and then to analyse the ways in which the local civil wars are being brought to an end.[1] Before doing so it will be useful to take a comparative look at an earlier period of state-generated violence in southern Sudan – the nineteenth century – to see if it suggests any patterns that can help us understand more recent events.

The Impact of the Nineteenth-century *Zariba* System

The majority of Dinka and Nuer first came into contact with the power of a centralizing state in the mid-nineteenth century, when the Egyptians opened up the tributaries of the White Nile and used them as a route for commercial and military penetration into southern Sudan. A state presence was established through a series of

---

1. An earlier version of this chapter, concentrating more on the political context of the Nuer civil wars, was presented at a seminar convened by the University of Copenhagen on 9–10 February 2001 and appears as Johnson (2001). It drew heavily on extracts from chapters 5, 6 and 8 in Johnson (2003), then unpublished. Johnson (2003) and now this chapter supersede Johnson (2001), though, inevitably, there is some repetition in all three publications.

armed camps (Arabic: *zariba* sing., *zara'ib* pl.) along the rivers or connected by caravan routes. Whether the camps were the headquarters of commercial companies or centres of government administration, they conformed to a common structure: composed of a fighting force recruited from various sources (both northern and southern Sudanese, free and slave), surrounded by captive or subject populations drawn from the immediate countryside, and often supported by local allies who participated in raids for cattle and slaves against their neighbours (Johnson 1992). As the *zariba* system was strongest in Bahr al-Ghazal, the Dinka were among those most severely affected by it: in fact the first *zariba* was established at Rumbek in Agar Dinka country in the mid-1850s; it was also the first major *zariba* to fall to a combined Dinka and Nuer force in 1883, as the Egyptian empire in Sudan collapsed.

The Nuer, for the most part, were remote from the active areas affected by the *zariba* system, but for a brief period, in the 1860s and 1870s, camps were established along the Zeraf valley, and the Nuer and Dinka of the region were drawn into the activities of the armed commercial companies. Among the Gaawar Nuer the merchants' main contact, Nuaar Mer, used the guns of the merchants' soldiers to enhance his own position, not only over neighbouring Dinka but over other Gaawar sections as well. By the time the merchants left the Bahr al-Zeraf in 1875 Nuaar Mer had amassed a large herd of captured cattle, but his involvement in raiding up and down the Zeraf valley had split the Gaawar apart, precipitating feuds even within his own primary section of Radh Gaawar. Nuaar Mer's immediate following, however, was not confined to his own section: it included Dinka as well as Gaawar, Lak and Thiang Nuer. In the general impoverishment Nuaar Mer helped to create, his own island of wealth attracted many of those he had so impoverished. The forces of opposition that eventually coalesced in his downfall in 1879 were also not confined to opposed Gaawar segments: while most came from the Bar primary section of Gaawar, many came from Nuaar Mer's Radh, some from the same Zeraf Nuer tribes represented in Nuaar's force and even some Dinka. Nuaar Mer's collaboration with the slavers had generated a number of feuds, but the coalition that defeated him did not conform to what is now considered the classic model of structural opposition (Johnson 1994: 128–41).

Very briefly, then, the events of the nineteenth century are not adequately explained by what is generally understood by the model of segmentary opposition. First, armed forces are notoriously mixed, and become even more so in the violence generated to maintain and recreate themselves. This was certainly the case in the *zara'ib*. Secondly, the commercial and military networks of the *zariba* system offered ambitious men new opportunities for accumulating cattle and followers that far exceeded those offered by existing kinship networks. Thirdly, the external intervention of the *zariba* system manipulated patterns of feuding, but it also disrupted and altered them and made possible other patterns of combination. These points should be borne in mind as we turn to the wars of the late twentieth century.

Events of the First Civil War and After

With a return to the use of tribal idioms to describe internal politics in Sudan, it is important to restate that neither the Nuer nor the Dinka are a 'tribe'. They are peoples who are subdivided into a number of different political entities known in both anthropological and administrative language as tribes. Historically, adjacent Nuer and Dinka tribes have often had more in common in their social and economic relations than they have had with more distant tribes of their own peoples. This had a distinct impact on political mobilization in the past and continues to do so in the present: one does not find in practice 'all Nuer' and 'all Dinka' mobilizing against each other (despite the stereotype propagated in some of the less reliable secondary anthropological literature, e.g., Kelly 1985). The principles of the pastoral 'common economy' operating in the flood-prone areas of the Sobat, Zeraf and Bahr al-Jebel valleys have meant that tribal boundaries are criss-crossed by networks of social relations (marriage, age-set membership, friendship), strengthened down the generations, complicating any attempt to identify discrete groups in opposition to each other (Johnson 1982, 1989).

Both the Nuer and the Dinka were slow to join the political disturbances in the south that immediately preceded and followed independence in 1956. As civil war spread throughout the southern provinces in the 1960s, the government in Khartoum responded by organizing and supplying semi-autonomous 'National Guards' to oppose the bands of Anyanya guerrillas infiltrating the countryside. Many Nuer were enthusiastic recruits into these National Guards, but, in being given licence to attack civilians as well as armed guerrillas, they soon became involved in raids against not just neighbouring Dinka but also other Nuer. With the government now involved in generating, rather than resolving, disputes, most Nuer east of the main Nile were driven into war against the government (Johnson 1994: 302–7).

Following the peace established by the Addis Ababa Agreement in 1972, the newly formed Southern Regional Government found that the resolution of intra-Nuer feuds presented as much of a difficulty as the resolution of inter-tribal fights. Part of the process of reinforcing the jurisdiction of the chiefs' courts, through which such disputes were resolved, was an acceleration of devolution in administration. This not only responded to but encouraged the drive for autonomy by many Nuer sections. Within Jonglei Province the Mor and Gun of the Lou Nuer, for instance, were allocated their own districts. Administrative devolution nationally decreed the division of all Sudanese provinces into two in 1976. Jonglei Province was detached from the Upper Nile, and the province boundary followed not a natural feature but the territorial boundary between the Lou and Jikany Nuer. Earlier colonial attempts to make provincial borders coincide with tribal boundaries had often increased, rather than diminished, conflict (Johnson 1982), and this new bit of boundary fixing was to have similar consequences.

International boundaries also interposed to impede the resolution of disputes between Nuer. The movement of the eastern Jikany Nuer into Ethiopia, which had been such a significant feature in the early twentieth century (Johnson 1986), was accelerated during the first civil war, as many Nuer sought refuge from government troops and Nuer Anyanya guerrillas set up their bases across the border. Even though many Nuer returned to Sudan, Nuer settlement in Ethiopia was encouraged by the Derg government in Ethiopia after the war. The Gaajak in particular established permanent settlements in Gambela region, displacing the Ethiopian Anuak there and becoming almost a separate tribe on its own, increasingly independent of the Gaajok and Gaagwang sections of Jikany. A number of Nuer were brought into local government in Gambela. With the Anuak being among the first to take up arms against the Derg in that area, the Gaajak influence was strengthened through their alliance with the Ethiopian government. The Nuer presence was further increased by the establishment of SPLA bases in Gambela region from 1983 to 1991, but the SPLA's own dominance in the area, supported as it also was by the Ethiopian government, had the effect of undermining the pre-eminent position the Gaajak had begun to develop for themselves.

The Beginning of the Second Civil War and Factionalism in the Movement

Having been slow to join the first civil war, the Nuer and Dinka Anyanya units of Upper Nile were among the most resistant to the Addis Ababa Agreement in 1972. It was initially opposed by Samuel Gai Tut, a senior Lou Nuer Anyanya commander, as well as by John Garang, a very junior Twic Dinka captain (Abel Alier 1990: 138). The Anyanya units of Upper Nile acquiesced only reluctantly, and for the first few years of the agreement ex-Anyanya resisted incorporation into the national army in a series of mutinies, the most serious being at Akobo in 1975, when the mutineers established new bases in Ethiopia. These became the nucleus of the 'Anyanya-2', guerrillas supported by the Derg in retaliation for Sudan's support of the Eritreans and other anti-Derg forces, such as the Oromo Liberation Front (OLF) and the Gambela People's Liberation Front (GPLF) of the Anuak. The Anyanya-2 were never a coherent or united force, but, as they became more active inside southern Sudan in the early 1980s, they became a nucleus for disaffected young men, increasingly disillusioned by President Nimeiri's attempts to weaken the Southern Region and his move towards imposing Islamic law. By 1983 there were numerous contacts between ex-Anyanya in the army, police and regional government and the Anyanya-2 operating out of Ethiopia and inside Upper Nile, Jonglei, Bahr al-Ghazal and Lakes provinces.

When the second civil war was ignited by the mutiny of southern troops in Bor in May 1983, a new fighting force, the Sudanese People's Liberation Army (SPLA), was formed in Ethiopia out of army mutineers and Anyanya-2 units operating mainly in Upper Nile and Jonglei. At the start it was composed mainly of Dinka and Nuer soldiers from those two provinces and its leadership reflected its composition, the most important early leaders all having served, at various levels, in the old Anyanya. The leadership struggle that accompanied the foundation of the movement was both ideological and generational, not tribal. The elder Anyanya veterans were in favour of reconstituting the old Anyanya, along with its platform of independence for the south, despite the structural weaknesses of the Anyanya and their failure to achieve their goal. But their patrons, the Derg, supported John Garang. His commitment to fighting for a united Sudan paralleled Ethiopian arguments against separatist movements within Ethiopia, and he was also the younger and better educated of the potential leaders, with a broader base of support among the exiled politicians who also joined the movement. Fighting between the new SPLA under the leadership of Garang and those Anyanya-2 units reluctant to join him was initiated by the Ethiopian army. Up through 1985 Nuer and Dinka were found in both factions.

The SPLA absorbed old Anyanya-2 units as it moved into new territories in western Upper Nile and the Bahr al-Ghazal Region. In western Upper Nile the local Dinka and Nuer had responded in two ways to the incursions of government-supplied Baggara Arab militias in 1982–84: some young men went to the SPLA training camp at Bilpam in Ethiopia and others, especially among the Bul and Leek Nuer, joined the resident Anyanya-2. In 1985 the mainly Bul Anyanya-2 around Mayom won a pyrrhic victory against the Baggara, losing most of their men in the fight. Thus it was that most Anyanya-2 units welcomed and joined the SPLA when they entered the area in force in 1984–85. One unit that remained aloof was a section of Bul Anyanya-2 under the command of Paulino Matip. Various reasons have been given for his refusal to join the SPLA, including his personal dislike for John Garang and his disapproval of SPLA-Nuer fighting along the Ethiopian border.

By this time the SPLA-Anyanya-2 combat had attracted Nimeiri's attention. Nimeiri, who needed a security arrangement that would enable Chevron to begin to exploit its fields in Upper Nile, hoped that he could raise a Nuer army to fight the 'Dinka' SPLA. In late 1984 and early 1985 contacts were arranged through D.K. Matthews, Nimeiri's Gaajak Nuer governor of Upper Nile Region, and William Abdallah Cuol, a Lak Anyanya commander. A new Anyanya-2 emerged, with William Abdallah Cuol as its leader, receiving arms, ammunition, uniforms and other supplies from the government.

William Abdallah Cuol moderated his platform of separation to federation, but inspiration for joining the government-backed Anyanya-2 had less to do with ideology than with individual motivation. The new Anyanya-2 drew recruits mainly from Bul Nuer under Paulino Matip, Lak Nuer with William Abdallah Cuol, a number of Lou Nuer sections from near Waat under Gordon Kong Banypiny (Mor Lou) and their neighbours the Jikany Nuer, including the Gaajak straddling the border with Ethiopia, under Gordon Kong Cuol. The Anyanya-2 did not enjoy the support of the Nuer as a whole or even the undivided support of the Bul, Lak, Lou or Jikany, and they continued to fight the Nuer in the SPLA. There was heavy fighting between the SPLA and the Anyanya-2 in Waat and Nasir districts. Anyanya-2 tactics made civilians in SPLA districts particular targets. The SPLA retaliated in a brutal war against the Gaajak Nuer, suspect because of their association with D.K. Matthews, and some Lou. Lou and Jikany Nuer civilians began killing individual SPLA soldiers for their weapons. Inside the SPLA internal security forces began arresting and executing Lou and Jikany soldiers suspected of dealing with the Anyanya-2. This in turn led to a spate of defections of Nuer SPLA to the Anyanya-2 (Nyaba 1997: 45–47).

The overthrow of Nimeiri in 1985, however, changed the pattern of personal alliances in the south, and some of those who had sided with him in the dissolution of the Southern Region in 1983 (such as D.K. Matthews) were now thrust out of power. The failure of Sadiq al-Mahdi's government to address any of the significant issues of the war meant that southern politicians who had previously been opposed to the SPLA began to move closer to it: it was, after all, the SPLA's military strength in the field that gave the southern parliamentary opposition some extra-parliamentary muscle. The SPLA, too, began to moderate its administrative practices. The defection of senior Nuer figures like D.K. Matthews to the SPLA and the death of William Abdallah Cuol (killed by his own men) enabled the SPLA to enter into a series of battlefield truces with many of the Anyanya-2 garrisons, especially those closest to the Ethiopian border. By 1988 Gordon Kong Cuol brought most of the Anyanya-2 into the SPLA. Those who stayed out of the new alliance included Paulino Matip's Bul Nuer, a garrison of Lak Nuer at New Fangak and a group of Lou based at Doleib Hill in Shilluk territory (Nyaba 1997: 48). But these garrisons effectively remained cut off from the countryside. Paulino Matip's harsh conscription policy and brutal treatment of his own soldiers led many Bul Nuer to defect to the SPLA under Riek Machar between 1988 and 1990 (Rone 2003).

This change in battlefield alliances had a significant impact on the SPLA's ability to keep its troops in Bahr al-Ghazal and the Nuba Mountains supplied and to shift troops to new theatres of war. In 1989 they captured Nasir and then rolled up many of the government's main garrisons east of the Nile. In 1990 Nuer soldiers played a significant role in the capture of the whole of Western Equatoria Province.

The Split in the SPLA and New Factionalism

The collapse of the Mengistu government in Ethiopia in May 1991 seriously reversed the SPLA's military momentum. The most immediate effect on the SPLA was its loss of protected bases, secure supply lines and sources of military as well as non-military supplies. The new Provisional Government of Ethiopia was not only hostile to the SPLA, because of its active involvement in the Derg's military operations against its opponents, but also had close links with the Sudanese army. There was a very real possibility that Ethiopia would allow the Sudan army to launch attacks on the SPLA from Ethiopian territory. In addition to this, fighting broke out between the Nuer and Anuak militias established under Mengistu, as the Anuak sought to re-establish their control over Gambela district. Groups of armed men, under nobody's direct control, became a very real threat to the quarter of a million Sudanese refugees inside Ethiopia.

The Sobat, Nasir, Waat, Akobo and Pibor areas had all suffered serious food shortfalls in 1988–90. They had been at the furthest end of the Operation Lifeline Sudan (OLS) supply line from East Africa and had received very little in the way of food, seeds, tools or fishing equipment. The local population of these areas therefore had no reserve margin of their own, and the agencies had not been able to establish a relief network. Thus both the local populace and the international relief community were unprepared to meet the sudden return of refugees that began in May 1991 (Johnson 1996).

The crisis caused by the collapse of Mengistu and the return of Sudanese refugees did not create the subsequent split in the SPLA, but it did affect the timing of the split and its subsequent direction. The systematic suppression of dissent within the SPLM/SPLA by its security forces had created considerable dissatisfaction with Garang's leadership among some military and civilian leaders. The fact that the two senior SPLA commanders at Nasir, Riek Machar (a Dok Nuer from the Western Nuer) and Lam Akol (a Shilluk), had quarrelled with Garang was apparently well known within the movement. But Riek and Lam were unable to persuade many of those similarly dissatisfied with Garang that he should be overthrown in a coup. The fact that Riek and Lam's plot was already an open secret meant that they could not back down. They were cut off from further supplies, saddled with a sudden influx of some 100,000 refugees and vulnerable to attack not only from the government's bases in Upper Nile but possibly even from Ethiopia.

With no strong internal support, but committed to some action as they were, the two Nasir commanders sought out the support of Nuer on both sides of the international border. Gaajak Nuer in the Ethiopian administration in Gambela (some of them originally from Sudan) faced an uncertain future with the advance of the OLF and GPLF and became strong supporters of the proposed coup. Approaches were also made to the Khartoum government via another Nuer, Lieutenant-Colonel Gaaluak Deng, governor of Upper Nile State and the commander of the Malakal garrison. In talks that lasted between July and August, an agreement was reached for supply and logistical support from the government and the transfer of the remaining Anyanya-2 units to the Nasir command (Nyaba 1997: 84, 89–90). Whatever the political or personal motivation of the two Nasir commanders, Khartoum saw this as an extension of its tribal militia strategy.

The announcement of the coup in August via SPLA two-way radios and over the BBC World Service did not bring the tidal wave of popular support the commanders either believed was coming or had promised their local supporters would come. The declaration of the total independence of southern Sudan as the new movement's goal was undermined by clandestine support from the government from which they were seeking to secede. This became clear in the fighting that followed. Government units of Anyanya-2 declaring for the Nasir commanders were involved in attacks on the Dinka civilians of Kongor and Bor districts between October and December 1991. The fact that the commanders soon lost control and their troops began indiscriminately killing civilians – a standard tactic employed by government militias in war zones both before and since – belied the political slogans emanating from Nasir. In order to stay alive as a military movement, the Nasir faction increasingly had to rely on government support and the mobilization of the Nuer. But the two could not be easily reconciled, and the movement became torn apart by the conflicting demands of the Nuer themselves.

Collaboration between the Nasir faction and Khartoum was publicly formalized with the agreement reached between Lam Akol and Ali al-Hajj Muhammad in Frankfurt on 25 January 1992. Following that agreement the military strategy of the Nasir faction was directed towards supporting government advances against the SPLA along the Equatorial front, even to the extent of dispatching troops there. This strategy was considerably enhanced by the defection in September 1992 of Garang's most senior Nuer commander, William Nyuon Bany, and the escape of some of Garang's political prisoners, including the Dinka commander Kerubino Kuanyin.[2]

The diversion of troops to the Equatorian front was not popular among many of the Nuer of the Sobat, especially the Jikany. Riek's public mobilization for an assault

---

2. For the government's involvement in securing Nyuon's defection, see Johnson (2003: chap. 8).

on Malakal early in 1992 had received popular support, and there was corresponding opposition when the troops were sent south instead. The ostensible reason for the Nasir faction establishing a presence in Eastern Equatoria was to open a route to the Ugandan border, where further supplies could be secured. Many questioned this at the time, since there was no indication that the Ugandan government was willing to supply them. Prevented by the SPLA from reaching the border, Nyuon contacted the Sudanese army in January 1993 and he and other commanders moved in and out of government garrisons (including Juba), establishing links with the Ugandan opposition Lord's Resistance Army and facilitating its incursions into Uganda. For the next two years Nyuon worked in close collaboration with the government, and troops, supplies and equipment passed between the government and Riek along the route opened up by Nyuon.

Parallel to the Eastern Equatorian strategy was the use of Kerubino Kuanyin to destabilize northern Bahr al-Ghazal in 1994–97. Kerubino advanced into Bahr al-Ghazal from western Upper Nile in June 1994 but was repulsed by the SPLA and forced to retreat to Abyei, in government territory. There he was reinforced and returned to Bahr al-Ghazal to continue his raids against civilian targets and relief centres. Over the next few years Kerubino, having failed to mobilize an anti-Garang Dinka force, increasingly relied on Paulino Matip and his Bul Nuer militia for supplies and recruits.

There was also serious fighting along the border between western Nuer and Tonj, Rumbek and Yirol Counties in Bahr al-Ghazal throughout 1994 and 1996. Civilians from both sides of the border were often mobilized, as the targets were usually opposing cattle camps or relief centres. But the heavy weapons used (including air support on the government side) and the level of logistical support and planning indicated that these were essentially military operations to which irregular civilian forces were recruited.

Renewal of the Nuer Civil War

With the failure of the Nasir faction to rally other parts of the south to its leadership, it increasingly had to rely on Nuer support to keep the movement viable. This was done through escalating Nuer–Dinka fighting along the border with SPLA-controlled territory and by supporting Jikany Nuer expansionist aims against the Anuak in the Gambela region of Ethiopia. This only widened fissures within Nuer society, resulting in the renewal of the Nuer civil war in the Jonglei area and along the Ethiopian border.

Many of the Nuer who rallied to the Nasir commanders in 1991 did so because they thought that now the Nuer would rule in the southern Sudan as the Dinka were accused of ruling before. The old antagonism against the Bor Dinka (extending to Garang's Kongor district) was revived, especially among the ex-Anyanya-2 Nuer, who lived at some distance from Kongor and Bor districts. The devastation of Kongor and Bor in 1991–92, resulting in the displacement of virtually the entire population of both districts, appears to have been initiated by the Fangak Anyanya-2. The Gaawar and Lou Nuer living directly on the border with the Dinka had many ties of kinship and affiliation that inhibited their participation in attacks on civilians. Lou Nuer of

Waat district were particularly upset by the attack on their Dinka neighbours and kin, especially because the Lou had come to depend on them for food and seeds throughout the previous three years.

War along the Nuer-Dinka frontier had a direct impact on security among the Nuer themselves. The serious inter-community fighting that broke out between the Lou and Gaawar in 1992 and the Lou and Jikany Nuer in 1993 was a direct consequence of the removal of a single SPLA administration. Prior to the split the whole of the area south of the Sobat had been placed under the administration of Bor. In the difficult years following 1988, the Lou Nuer of Waat district had benefited from access to their Dinka neighbours in Kongor and Bor and from regulated access to the Sobat dry-season grazing grounds and fishing pools, which they shared with the Jikany. Following the split, the Jikany asserted their exclusive claims to the Sobat, and Riek, dependent on their goodwill as he was, recognized these. At the same time, the UN relief effort focused on the Sobat around Nasir. The split in the SPLA cut the overland relief route from Bor to Ayod and Waat just when severe late rains virtually destroyed the 1991 harvest. In 1992 some Lou raiding parties attacked the Gaawar for cattle as the nearest source of food now available to them.

By the end of 1992 the Lou Nuer had numerous grievances: they felt that they had not been given sufficient attention in the relief effort and that they were also being excluded from necessary riverain pastures and fishing grounds. These grievances were presented to Riek personally at the end of 1992, but he took no action. The Jikany, for their part, now no longer had easy access to the Gambela region. Riek's forces had supported Jikany assaults on Anuak villages in the Itang area in January and July 1992, but these had been decisively repulsed by Ethiopian troops (Kurimoto 1998). One effect of this ill-judged action was to eliminate any possibility of relief supplies coming to the Sobat region from Ethiopia. Competition for control of the Sobat was sharpened.

Fighting between Lou and Jikany groups broke out in the dry-season pastures in 1993 (Kwacakworo 1994), just at the same time that heavy fighting broke out between the two SPLA factions in Kongor along the Lou border with the Dinka. Subsequent to this fighting, the Lou also complained that they had received very little support from forces in neighbouring Ayod and Nasir areas when Garang's forces counter-attacked as far as Waat. In the immediate aftermath of the SPLA counter-attack, various units of Nuer SPLA-United began fighting each other.

SPLA-United commanders from both the Jikany and Lou now approached the government in Malakal seeking arms, ostensibly with which to fight Garang. These were given freely, and with them the commanders armed groups of their fellow citizens. Throughout the early part of 1994 there were clashes between the Lou of Waat and the Jikany of Nasir, leading to the destruction of Ulang and Nasir itself. Over 1,200 persons were killed in this fighting. Pressure from concerned Nuer exiles led to the convening of a large inter-Nuer peace conference in Akobo during the 1994 rainy season, where an attempt was made to sort out inter-Nuer problems and improve relations with the Anuak in Ethiopia (Duany 1994).

The peace conference, which brought in respected mediators from other parts of Nuerland, achieved one goal, which was to set out the terms by which the main antagonists would agree to stop fighting and the compromise each was willing to accept. It also revealed the extent to which individual commanders within SPLA-United had collaborated with the government and, in doing so, reinforced the momentum for repudiating that collaboration and reaffirming the stated goals of the movement. For all that, it failed to achieve its aims because, in the end, Riek failed to institute those measures necessary to implement the agreement. Troops were not sent into the disputed pastures to keep the peace, and those commanders previously involved in the fighting were not reassured that they would not face retaliation. The fighting that broke out again between armed units lasted throughout 1995 and contributed to the final disintegration and demise of the movement.

In 1995 Nuer soldiers stationed near the Uganda border at Jebel Lafon re-defected to the SPLA, taking their commander, William Nyuon, with them. Reunification talks between the SPLA and SSIM began and, while some soldiers and leaders returned to the fold, many more did not. Riek was put under considerable pressure by leaders from his own area (Dok) to maintain the integrity of a Nuer-led movement. But by the end of 1995 that movement was a virtual movement only, lacking any real substance or coherence. With different parts of his army now divided between the SPLA and SSIM and fighting each other, Riek formally signed a Peace Charter with the government in Khartoum on 10 April 1996.

Riek proclaimed that this agreement would give the south its independence. His representatives in East Africa gave it a more restricted interpretation. As one explained to an OLS official, SSIM considered itself in a state of ceasefire with the government and at war with the SPLM.

> This is justified by the explanation that SSIM finds it impossible to be at war with two parties at once and sees John Garang as a greater threat to the survival of the Nuer people than is the Government ... One thing is clear: the SSIM and the Nuer people feel a strong sense of persecution. They insist that the international community (USA, Ethiopia, Uganda, etc) as well as the SPLM is seeking to wipe out the Nuer as a political force. (Levine 1996)

Far from affirming the movement's commitment to independence of the whole south, this agreement was a tactic to ensure the movement's dominance among the Nuer. Coinciding with his move to Khartoum, Riek mobilized Western Nuer forces to attack Lou Nuer SPLA in Jonglei.

The 1996 Peace Charter was followed a year later by a formal Peace Agreement, which amalgamated all southern militias into a new South Sudan Defence Force (SSDF), ostensibly under Riek's command. But, far from providing the Nuer with either political or military unity, the implementation of both the Peace Charter and the Peace Agreement opened up a new front in the war, centred on the oilfields inside the Nuer homeland.

War in the Oilfields

Nothing demonstrated the falsity of the internal peace more than the continuation and expansion of the Nuer civil war. It was, in fact, the internal peace which added fuel to that civil war. Prior to 1996 it had been confined mainly to factional fighting within Riek's command in Upper Nile and Jonglei. After 1997 its focus shifted to the Western Nuer in Unity State. Here, the power play between two Western Nuer rivals, Riek Machar and Paulino Matip, was fuelled by the very government to which they were both allied. The fostering of the civil war among the Western Nuer, while intensifying insecurity in what was supposed to be a government-held area, served the government's purpose in that it neutralized the southern guerrilla factions best positioned to interfere with the exploitation of the Bentiu oilfields. Matip, who had worked closely with President Bashir since their days in Mayom together, could be counted on to do the government's bidding in the field.

Paulino Matip was as near a reincarnation of an old *zariba* captain as it was possible to get.[3] Using his Bul Nuer territory and its proximity to both Kordofan and Bahr al-Ghazal, he built up a small trading empire dealing in sorghum – which was grown locally – and cattle – which were either looted directly from the Dinka or brought in from the cattle market in Ler. He maintained an autonomous military and economic base, despite his nominal inclusion in the Nasir faction after 1991. In fact, the split in the SPLA opened up new economic opportunities for him when his headquarters at Rupnaygai became a major trading entrepôt and the gateway by which Arab merchants approached the Western Nuer, bringing in much needed currency and manufactured goods and taking out cattle for markets in the north. Paulino thus controlled access to the wealth that was needed to revive the rural economy of the Western Nuer. As a Major-General in the Sudanese army he also had direct access to military supplies. Riek's appointment of Paulino as the SSIM governor of its Liec State in 1994 was merely a recognition of the actual state of affairs.

Riek had more need of Paulino than Paulino of Riek, and Riek's attempt to put his own man in the governorship of western Upper Nile after returning to Khartoum led to an open breach, with Paulino's troops taking and sacking Ler, Riek's old headquarters and the centre of his relief operations, in July 1998. There were shoot-outs between Riek's and Paulino's forces in Khartoum and further inter-Nuer fighting within the SSDF in Juba.

Riek's protests at the government's continuing support for Paulino had no lasting effect (Human Rights Watch 1999: Appendix F). The intensification of oil exploitation in the Bentiu oilfields in 1998–99 brought with it an increase of Sudan army and other armed personnel as well as foreign oil workers into the region, contributing directly to another outbreak of fighting between a profusion of Nuer factions beginning in April 1999. Government policy required the clearance of any civilian population from around the oil installations, and Matip's forces were the main instrument of this clearance. Large areas in and around the oilfields were

---

3. The re-emergence of other aspects of the *zariba* system can be seen in the style of construction of forts protecting the oil installations (Gagnon and Ryle 2001).

depopulated as Nuer civilians fled to Bahr al-Ghazal seeking SPLA protection, with Paulino receiving support from government Popular Defence Force units and Arab militia on the ground and Khartoum's Antonov bombers and helicopter gunships in the air. At least two groups of Nuer militia began gravitating towards the SPLA. Riek's troops steadily abandoned the state.

The internal peace encouraged the creation of named southern groups, in order to give the appearance of embracing the plurality of southern political opinion. The proliferation of initials continued as the peace unravelled. Paulino Matip renamed his force the South Sudan United Army (SSUA) when he broke away from Riek's SSDF in 1998, but the former initials were replicated in other breakaway factions: an SSDF-2 in Juba, an SSDF-United and an SSDF-Friendly-to-the-SPLA among the Western Nuer. Riek added yet another set of initials in 2000, with the renaming of his remaining coalition of commanders the Sudan People's Defence Force (SPDF).

With chaos ramifying throughout the Western Nuer heartland, the SPLA found itself in the unusual position of protecting refugee Nuer. This was the context in which an intertribal conference, sponsored by the New Sudan Council of Churches and guaranteed by the SPLA, was held at Wunlit between groups of Western Nuer and groups of Dinka from Tonj, Rumbek and Yirol, leading to a peace settlement between them in March 1999 and an agreement by the SPLA not to sanction cross-border raiding (see Hutchinson, this volume). Local truces between SPLA and SSDF forces in Jonglei had held throughout much of 1998 and 1999, with some Nuer forces defecting to the SPLA outright. Some of the same groups and persons who had participated in the Nuer–Dinka reconciliation meeting subsequently organized a large meeting of Nuer in Waat, drawing on people from both the SPLM/A and UDSF/SSDF factions. In December 1999 they announced the formal break of the majority of the Nuer with the government and their willingness to renew cooperation with the SPLA.

Riek's political organization in Khartoum had had no part in the Wunlit conference and was left stranded by the decision in Waat. Without an army, without followers, Riek was now without a movement. He tried to regain the initiative by leaving Khartoum in December 1999 and resigning from the government two months later, but, lacking any real support in his home area, he had to leave western Upper Nile for Nairobi, where he tried to maintain the appearance of the leader of a viable political movement.

Having formally extracted themselves from the alliance with Khartoum, the organizers of the Waat conference were divided over their future strategy. Some SSDF commanders were still not eager to go back to the SPLA. They put pressure on Michael Wal Duany, one of the principal Nuer organizers of the Wunlit and Waat meetings, to form a new movement, on the basis of which, they argued, they could then gain external support for arms and supplies. Duany proclaimed the formation of yet another initialled movement: the South Sudan Liberation Movement (SSLM). This did not have the desired effect: external support was not forthcoming, the SPLA remained suspicious of the new movement's intentions, some Western Nuer SSDF commanders moved back to cooperating with the SPLA or the SPDF, and other commanders who had not been part of the Waat conference announced their intention of continuing to fight on the side of the government.

It has been difficult to keep track of the post-2000 alliances between Nuer militias or even to have a clear idea of their objectives.[4] A merger between the SPDF and SPLA was announced in May 2001, only to be repudiated by the SPDF a few days later. With the changed international climate following the 11 September attack in the US and the momentum towards possible negotiations in Sudan generated by the Danforth initiative at the end of 2001, Garang and Riek finally announced their reconciliation on 7 January 2002 and the re-amalgamation of the SPLA and SPDF. Some SPDF commanders based along the Sobat and in the Gaajak and Lou areas then announced their defections from the SPDF to the SSLM. The SSLM, while remaining committed to peaceful collaboration with the SPLA, resisted reincorporation into the SPLA under Riek's command. In Western Nuer Peter Gatdiet of the SPLA and Peter Paar of the SPDF formally merged in February 2002 in the face of a renewed government offensive around the oilfields.

There is very little hard evidence emerging from the field concerning the recruiting patterns of the different militias or the motivations of the ordinary soldiers, as opposed to their commanders. What follows here is necessarily tentative.

It is not really possible to identify alliances according to the tribes or even primary sections of the commanders. Among the Western Nuer Paulino Matip and his former second-in-command, Peter Gadiet, are both Bul Nuer from the Kuac primary section. Two of Riek Machar's other commanders, Tito Biel Cuol and Peter Paar Yak, are, like Riek, Dok Nuer. Both Matip and Gatdiet were involved in fighting soldiers commanded by Tito Biel and Peter Paar in 1997 and 1998. In 1999 Gatdiet broke with Matip over the oil issue and took most of the Bul soldiers with him. Both Tito Biel and Peter Paar received supplies from the SPLA in their continued fighting with Matip in 1999. Gatdiet and Biel at first aligned themselves with the SSLM in Waat in November 1999, but Biel rejoined Riek Machar after the latter resigned from the government in 2000. Gatdiet, now reabsorbed into the SPLA, fought Riek's troops in 2000, but for a while assisted Paar's troops. When Gatdiet's and Paar's soldiers began fighting each other in mid-2000, Gatdiet, a Bul Nuer, was supplied by the SPLA while Paar received government supplies through Matip (Bul), though still fighting against the government. Biel and Paar, both Dok, came into conflict and Riek had to separate them, sending Biel to help organize troops among the eastern Jikany in Maiwut. Riek was not safe in his own homeland and in 2000 made the eastern Jikany area his base of operations. When Riek was reconciled with Garang in 2002, Biel was among those of his commanders who rejoined the government and Matip, as later did Gatdiet.

It is also difficult to make a clear identification of militias by tribal allegiance. It has been common for external reporters to refer to Matip's army as a Bul Nuer militia. His original support came from the Bul, and he often supplied Bul Nuer troops to Kerubino, who was ostensibly the commander of a Dinka army, but who

---

4. The most detailed account so far is Rone (2003).

was also Matip's father-in-law. But Matip had difficulty keeping his Bul troops. Many defected to the SPLA in 1990, before the split. More went over to Gatdiet when he defected in 1999. Matip was subsequently reinforced directly from Khartoum – mainly southern Sudanese of any origin conscripted by the government from the shanty towns around Khartoum and sent south. Those who later deserted to the SPLA include Lou Nuer and Shilluk. So Matip's army, while well supplied by the government, lost any clear 'ethnic' identity.

During the fighting between Matip and Riek's forces in 1997–98, prisoners of war were often pressed into service by their captors, regardless of their tribal origin. Some groups, such as the Leek living along the Bahr al-Ghazal, are said to have supported whichever faction was occupying their territory at the time. Prior to 2002 Gatdiet's soldiers were reported to be joining whichever commander was best supplied, whether by the SPLA or the government.

There was a broader division in Western Nuer between those living north of the Bahr al-Ghazal (Bul, Leek, Jikany) and those living south (Jagei, Dok, Nyuong). There had been conflicts between the Leek and Bul Nuer, but before 2002 the Leek and Jikany militia tended to side with Gatdiet's Bul. Before the 1991 split in the SPLA, Matip's Bul troops generally held Riek's Dok accountable for the losses they suffered and, though the two tribes do not border each other, a Bul–Dok animosity was one of the factors that drove Gatdiet's and Paar's troops apart in 2000. But, with the expansion of fighting between Gatdiet and Paar to the Nyuong Nuer area, tensions also arose within Paar's command. Early in May 2001, there was a shoot-out between some of his Dok and Nyuong soldiers in Nyal (Nyuong Nuer territory), with the Nyuong accusing the Dok of bringing their quarrel with the Bul to the Nyuong.

A different situation pertained to the area east of the Bahr al-Jebel, where the government maintained militias among the eastern Jikany (Gordon Kong Cuol), the Lou (Simon Gatwic) and the Lak (Gabriel Tang-Ginya), eventually placing them with all its other southern militias under one command. The eastern Nuer militias sometimes increased 'recruitment' through forced conscription drives during military offensives in territory outside their control (CPMT 2003). These were also directed against other southern peoples, most notably when Gabriel Tang-Ginya's troops were used to sweep through the southern Shilluk kingdom (CPMT 2004).

Even though the SSLM formed at Waat was unable to maintain a military coalition, the level of fighting in the area declined because of the unwillingness of local people to lend their support to the militias. Many Lou, for instance, joined Simon Gatwic only to get arms and ammunition with which to protect their homes, after which they left him. The Gaagwang section of the eastern Jikany required all militiamen, whatever their affiliation, to store their arms on entering Gaagwang territory. It was mainly in Ethiopia, in Gambela region, that armed Jikany Nuer allegedly affiliated with either the Sudan government or Riek Machar used their military strength to actively continue their displacement of Anuak in local politics (Kurimoto 2002: 236–38). Before the government renewed militia activity in the area in 2004, SPLA and SSLM soldiers and Lou and Jikany civilians freely associated with each other in Akobo, the site of the only functioning hospital in the region after the hospital in Nasir was destroyed in the Lou–Jikany war.

Post-2005 Events and Conclusion

The signing of the Comprehensive Peace Agreement (CPA) and formation of an SPLM-led Government of South Sudan (GOSS) in 2005 dramatically altered the political landscape throughout the Southern Sudan. By the terms of the CPA all 'Other Armed Groups' (OAGs), militias aligned either with Khartoum or the SPLA, were to be integrated either into the Sudan Armed Forces (SAF) or the SPLA, or disarmed and demobilized. Following the death the SPLM/SPLA leader John Garang in a helicopter crash in July 2005, his successor, Salva Kiir Mayardit, accelerated the process of wooing the Khartoum-backed Nuer militias, with the result that most, including Paulino Matip and Simon Gatwic, chose to be integrated into the SPLA, leaving only a few holdouts maintaining their connections with Khartoum's State Security and SAF. In addition to this the new GOSS in Juba determined on a general programme of disarming civilians who were not part of any formal armed group, beginning the process of disarmament among the Nuer of Jonglei State, particularly the Lou and Gaawar (Young 2006, 2007; Garfield 2007).[5]

There are a few brief points to emphasize in conclusion. First, to point out the obvious, the scale of military operations moved beyond the level of tribal war. This is as true of Nuer-Anuak fighting in Ethiopia, Nuer-Dinka fighting in the Jonglei area and Nuer-Dinka raiding along the Upper Nile–Bahr al-Ghazal border as it is of the long-standing Baggara Arab raids against the Dinka. The logistical and organizational support given by the different armies in the region, the involvement of uniformed soldiers and the deployment of advanced weaponry, including rocket-propelled grenades, heavy machine guns, artillery and (on the government side only) air support, are evidence that each army directed the fighting for its own strategic aims. This pattern of military organization was later replicated in Darfur (Amnesty International 2004; Human Rights Watch 2004). Secondly, the failure of the Nasir faction to establish its national credentials led directly to its creation of a Nuer army as the base of its support. Thirdly, military mobilization along tribal boundaries made use of local tensions and existing patterns of fighting, but these tensions did not, by themselves, lead to fighting. It was when political organizations, whether the Sudanese government or the guerrilla administrations, frustrated those local institutions set up to contain and resolve tensions with the intention of encouraging fighting that fighting has broken out. Fourthly, cutting the Nuer off from their neighbours in the attempt to mobilize a Nuer army and the reliance on government supplies to that army led directly to a series of civil wars among the Nuer themselves. Fifthly, the longer fighting continued and the greater the need for armed groups to maintain themselves through forced recruitment, the more mixed the armies became. Paulino Matip is the nearest we have in recent times to a nineteenth-century

---

5. Reports by various NGOs on the transformation of the OAGs and the disarmament of civilians unfortunately provide very little ethnographic detail and also display a shallow historical understanding of the groups involved. We still do not have detailed and reliable information about the sectional or generational patterns of militia recruitment, the reasons for their fragmentation, or motivation for rejoining or holding out against the SPLA.

zariba commander, and his forces came to resemble the composition of those old *zariba* armies, with soldiers of different origins, both free and coerced.

Finally, there is as long a history of local patterns of reconciliation as there is of fighting, and these patterns of reconciliation have begun to reassert themselves with the encouragement of local administrations and international organizations. Popular support for these efforts has had an inhibiting effect on the activities of local commanders and their troops.

Chapter 2

Peace and Puzzlement: Grass-roots Peace Initiatives between the Nuer and Dinka of South Sudan

Sharon Elaine Hutchinson

> There is nothing in the hearts of Nuer and Dinka to make this conflict.
> Something happened which is beyond our powers.

A ferocious white bull, tethered only at the neck, wrestled for its life. Kicking and groaning, the great 'white one', Mabior, nearly broke free from a tightening circle of hundreds of shouting and dodging Dinka and Nuer spiritual leaders and chiefs. 'You are fierce like the conflict of our people is fierce. You are the wild one and you will accept this peace!' Finally, someone caught the bull's tail and the crowd closed in. Heaving in unison, the men forced the bull on its side and held its legs. In a gesture of respect, some Dinka women rushed forward to cover the bull's genitals with a cloth skirt. Still writhing and moaning, the bull's head was pushed to the ground and its throat cut. The crowd immediately burst into song, confident that Mabior's blood would cleanse them of all the evils of the previous eight years.[1]

The date was 27 February 1999, the place Wunlit, south Sudan – a freshly constructed set of mud-and-thatch shelters occupying a parched and isolated stretch of savannah in the eastern Bahr el-Ghazal Province. More than 300 Nuer and Dinka chiefs and spiritual leaders had gathered, along with a smattering of southern politicians, military observers, expatriate church workers and several thousands of supportive civilians, in the hope of ending the militarized and ethnicized south-on-south violence that had decimated their homelands since 1991. That was the year in which the Sudan People's Liberation Army (SPLA), which has been fighting to overthrow the northern-dominated, national, Islamic state in Khartoum since 1983, split abruptly into two warring southern factions. This split began as an internal power struggle between Dr John Garang de Mabior, the SPLA Commander-in-Chief and a Bor Dinka, and Dr Riek Machar Teny, a Dok Nuer, who formed a breakaway

---

1. Field research for this chapter was made possible by the generous support of the Harry F. Guggenheim Foundation and the Pew Charitable Trusts, for which I am deeply grateful.

rebel faction with two other high-ranking SPLA officers, following their botched coup against Garang. When neither side managed to crush the other, the violence swiftly escalated into a full-scale military confrontation between the Dinka and Nuer, the two largest ethnic groups in the south, which formed the bulk of the SPLA's fighting force. Swinging their guns away from their common enemy in the north, Nuer forces allied with Riek Machar and Dinka forces loyal to John Garang began attacking and looting each other's rural constituencies. Using 'scorched-earth' military tactics against a civilian population armed with little more than spears, rival southern field commanders and warlords struggled to create their own spheres of political and economic dominance, their own ethnic security maps.[2]

Whatever remnants of law and order existed in Nuer and Dinka communities before the SPLA split soon dissolved in a sea of southern blood, most of which flowed from the most vulnerable segments of the civilian population. Local ethical codes of warfare, which had been respected by generations of Nuer and Dinka combatants, were simply cast aside. Suddenly everyone became vulnerable to attack, even newborn children, pregnant women and the elderly. Widespread recourse to rape, torture and abductions provoked massive population displacements and, inevitably, a region-wide famine. Having had their cattle looted, their crops slashed and homes burnt, many Dinka and Nuer families sank into previously unimaginable states of poverty. Although rural Nuer and Dinka civilians knew that this fratricidal violence was not in their interests, they felt powerless to stop it owing to the unchecked predations and abuses of rival southern field commanders and warlords.

The 1990s were a decade of internal power struggles, inter- and intra-factional fighting, secret alliance building (including with the government), instability and violence. Capitalizing upon the chaos, the government of Sudan intensified its bombardment of Dinka and Nuer villages in the northern Bahr el-Ghazal and Western Upper Nile, as well as periodically banning all relief flights into the area. Backed up by thousands of northern Baqqara[3] raiders and several government-allied Nuer militias, the National Islamic Front Government in Khartoum endeavoured to reassert military control over the vast oil wealth lying beneath Nuer and Dinka regions of the Western Upper Nile (Verney 1999; Amnesty International 2000; Gagnon and Ryle 2001; Rone 2003).

Finally, in June 1999, a major political breakthrough occurred: a small group of Nuer and Dinka chiefs bravely embarked on a grass-roots peace initiative of their own. With financial and logistical assistance from the New Sudan Council of Churches (NSCC), eight leading Dinka and Nuer chiefs were plucked from their rural homelands and flown to Lokichokkio, Kenya – the operational headquarters of

---

2. The bloody aftermath of the 1991 collapse of southern military unity has been documented in depressing detail by a number of social scientists and human rights workers. Apart from several recent articles of my own (Jok and Hutchinson 1999; Hutchinson 2000, 2001; Hutchinson and Jok 2002) incisive historical accounts may be found in Nyaba (1997), Johnson (1998, 2001) and Lesch (1998), as well as two excellent volumes published by Human Rights Watch (1999 and Rone 2003). See also Johnson, this volume.
3. Baggara and Baqqara (see Johnson, this volume) are alternative spellings of the same name.

Operation Lifeline Sudan, a UN-led consortium of international relief agencies operating in the region since 1989. Freed from the seven-year-long stranglehold on face-to-face communications imposed by rival southern military factions, these Dinka and Nuer chiefs were offered a forum in which to air their grievances, share their insights into the root causes of the conflict and, most importantly, begin negotiating a civilian-led truce that would stem the bloodshed. Other participants in the 'Dinka and Nuer Chiefs Peace Workshop', held in Lokichokkio from 2 to 10 June 1998, included a dozen or so Dinka and Nuer church leaders, two leading Dinka lawyers and a handful of south Sudanese, Kenyan and American NSCC organizers. All southern military personnel were barred from attending. This prohibition undoubtedly contributed to the remarkably frank, freewheeling and constructive discussions that took place. Within a day or two, it became apparent that a subset of Nuer and Dinka chiefs hailing from the west bank of the Nile was making especially rapid progress towards a grass-roots peace agreement.[4] Consequently, before the workshop adjourned on 10 June 1998, western Nuer and Dinka chiefs resolved to meet again, this time in south Sudan, for a more comprehensive peace conference to be held during the coming dry season.

This resolve culminated in the 'Dinka and Nuer Peace Conference', held in Wunlit, Bahr el-Ghazal, between 27 February and 8 March 1999. Nevertheless, were it not for the personal support for the conference's objectives offered by Commander Salva Kiir Mayardit, the second highest-ranking SPLA officer after Garang and a Bahr el-Ghazal Dinka, the Wunlit Peace Conference would not have taken place. John Garang did not publicly endorse this conference in any way. Moreover, less than a week before the conference was scheduled to begin, a party of Dinka raiders attacked several Nuer cattle camps. These raids resulted in nine civilian deaths, the abduction of several women and children and the loss of some 2,000 Nuer cattle. When the Nuer chiefs most directly concerned were asked by the NSCC organizers whether or not they wanted to suspend the conference in light of these events, the chiefs unanimously affirmed their willingness to continue working for peace. This affirmation required considerable courage since it meant that these chiefs would have to travel deep into Garang's SPLA-Mainstream territory without any military escort. It also put pressure on SPLA leaders – and particularly on Commander Salva Kiir – to identify and punish the aggressors, which he did.

In contrast to Garang's resounding silence, Machar – who by that time had signed a separate peace agreement with the Khartoum government – wrote a formal letter to the NSCC organizers in which he expressed his unconditional support for the conference's aims. A delegation of Nuer politicians who joined the government together with Machar in 1997 also flew down from Khartoum to attend. Their uninvited attendance, however, proved aggravating to some attending chiefs and SPLA-Mainstream representatives, who expressed fears that the Nuer Khartoum delegation would seek to 'hijack' the peace process for its own propagandist purposes.

---

4. The failure of the NSCC organizers to secure the attendance of any Dinka chiefs hailing from the east bank of the Nile foreclosed any possibility of real progress being made on that front at the Lokichokkio conference.

Despite these smouldering tensions, the 1999 Wunlit Peace Conference represented a major step forward for Nuer and Dinka communities from the Western Upper Nile and Bahr el-Ghazal Provinces. Although more ceremonial and reserved than the Lokichokkio Peace Workshop, the Wunlit Peace Conference largely succeeded in restoring an atmosphere of inter-ethnic trust between Nuer and Dinka communities west of the Nile. Conference participants declared a general amnesty for all crimes committed before 1 January 1999 and agreed either to return all abductees or, in the case of women who expressed a desire to remain in their new homes, to formalize their marriages with the payment of bridewealth cattle. Border courts and police, armed with two-way radios, were to be re-established and villagers encouraged to repopulate the vast 50–100 kilometre-wide stretch of rich grazing grounds and agricultural lands abandoned during the height of the crisis. The response was immediate. Nuer and Dinka civilians began circulating freely throughout the region, re-establishing trade linkages as well as opening up informal channels of communication between opposed southern military units and field commanders. Cross-border cattle raiding also declined dramatically.

However, in August 2001, the military situation on the Nile's west bank remained extremely tense. The government of Sudan, feeling deeply threatened by the spirit of regional peace and reconciliation fostered at Wunlit, proceeded to intensify its assaults, both airborne and ground-based, on Nuer and Dinka civilian populations located in the 'oil zone'. Declaring the Wunlit Peace Agreement a 'conspiracy' fostered by American-based religious organizations and a national security threat, the central government launched a massive recruitment drive immediately after the Wunlit Peace Conference in order to expand its control over strategic oilfields in the Western Upper Nile[5] as well as to preclude any further progress towards a reconciliation of Nuer and Dinka civilian and military leaders. For example, a spin-off south-on-south peace conference held in Wichok, Western Upper Nile, during October 1999, which sought to reunite the Bul Nuer, was abandoned when the government bombed Wichok and other Nuer settlements in the area. A third 'People-to-People Peace Conference' held in Liliir (Lirlir), south Sudan, during May 2000 was intended to extend the spirit of the Wunlit agreement to Nuer, Dinka, Anyuak,[6] Murle and other communities east of the Nile. Unfortunately, this event failed in its objectives and, indeed, appeared to have intensified political rivalries among some Nuer groups.

Nevertheless, the spirit of inter-ethnic peace and reconciliation fostered between Dinka and Nuer communities on the west bank continues to hold, despite recurrent

---

5. These attacks cleared the way for the government's completion of the first phase of a 1.6 billion dollar oil development scheme, funded by a consortium of international petroleum companies led by the Canadian giant, Talisman. The scheme involved the completion of a massive oil pipeline running directly from the northern Western Upper Nile to newly constructed refineries and export stations in the far north. For more information on the pipeline, which became operational in August 1999, see Verney (1999), Amnesty International (2000) and Human Rights Watch (2003).

6. Alternative spelling of the Anywaa of Dereje's and Falge's chapters, this volume, and the Anuak of Johnson's and James's chapters, this volume.

surges in intra-Nuer factional fighting and despite concerted government attempts to undermine it. Indeed, the Wunlit Agreement has already proved vitally important to the estimated 70,000 Nuer and Dinka civilians who were driven from the oilfields in the Western Upper Nile during 2000 by a lethal mixture of northern troops, Baqqara Arab militia and pro-government Bul Nuer forces led by Paulino Matiep Nhial. Fortunately, most of these civilians eventually found refuge in Dinka communities of the eastern Bahr el-Ghazal, where they were hospitably received, thanks to the spirit of inter-ethnic trust and harmony forged at Wunlit.

Having been fortunate enough to attend both the 1999 Wunlit Peace Conference and the 1998 Peace Workshop in Lokichokkio, I was moved by the sincere efforts of participants to develop a broader understanding of the violent forces shaping their daily lives. In a war as twisted, lengthy and lethal as that in south Sudan, it has not always been easy for ordinary people to identify the principal lines of conflict – to know precisely who is fighting against whom and why. Caught off guard by the violent turn of events triggered by the 1991 splitting of the SPLA, most Dinka and Nuer men, women and children were simply swept up in an expanding spiral of southern military raids and counter-raids not of their own making. 'We don't [even] know why we are fighting!' exclaimed one bewildered Nuer chief in Lokichokkio. 'Why are we [the Nuer and Dinka] killing our women and children?' But reaching a consensus about the dynamic forces fuelling this fratricidal violence proved far more difficult for conference participants. Everyone recognized that this violence had little to do with the daring, cross-border, cattle raids staged by generations of Nuer and Dinka youths seeking to demonstrate their courage and fighting prowess. Local patterns of competition over cattle and other scarce economic resources were completely corrupted once Garang and Machar began to fight out their political differences along ethnic lines. It had become a soldier's war in both equipment and purpose.

The full weight of this human tragedy has yet to be measured, and it is not my intention to do so here. Rather, I wish to focus in this chapter on the remarkable range of historical and sociological explanations for this calamity offered by leading Dinka and Nuer chiefs and religious leaders who attended one or both of these grass-roots peace conferences. As an anthropologist who has struggled to keep pace with the rapidly shifting political and military currents of south Sudan, I was struck by the diversity of indigenous theories and opinions expressed at these conferences; it is this diversity that I have sought to capture and convey here.[7]

---

7. The quotations that follow have been gleaned from personal tape recordings I made during both meetings as well preliminary records of these events gathered on site. All translations from the Nuer language are my own. For translations of Dinka speeches, I have relied on recorded, simultaneous translations into Nuer and English provided at both events. Although both conferences were vital to the restoration of inter-ethnic trust, it was my impression that the presence of military observers at Wunlit inhibited some civilian leaders from airing the full weight of their grievances, particularly grievances against southern military personnel. The more ceremonious and platitudinous tenor of that event contrasted sharply with the far more raw and unguarded speeches voiced at the unmonitored workshop in Lokichokkio. Consequently, the reader should take careful note of the varying contexts of the quotations cited, most of which date back to the 1998 Lokichokkio conference.

First Hypothesis: 'This is not a fight between the Nuer and the Dinka'

Organizationally, both conferences foregrounded ethnicity as the principal axis of violent conflict. The NSCC organizers struggled – with varying success – to secure a balanced representation of Nuer and Dinka chiefs and religious leaders from diverse areas at each event. Seating arrangements, particularly at the Wunlit Conference, reinforced this ethnic polarization. And yet, on both occasions, many Nuer and Dinka chiefs and religious leaders adamantly rejected the premise that the conflict was ethnically based. For example, in a revealing moment towards the close of the Wunlit Conference, two Bul Nuer chiefs arrived late and breathless, having traversed more than 200 kilometres on foot in the hope of addressing the assembly before it disbanded. In their opening remarks, one of them voiced his surprise at finding the seating arrangements ethnically polarized. He had expected that attending Nuer and Dinka chiefs would be seated together on one side and all military personnel on the other. Since he belonged to a splinter group of Bul Nuer that had recently joined forces with the SPLA, he was at first uncertain where to stand but eventually moved to the Dinka side of the meeting hall.

One Dinka chief present in Lokichokkio also rejected the idea that the violence was motivated by ethnic animosity. In his words:

This [post-1991] conflict began when people were one. This conflict did not start because there were Nuer people who went to the Dinka area and stole cattle or Dinka people who went to the Nuer area and stole cattle. It happened suddenly and without warning. What happened when the conflict started was that the Dinka who were living in the Nuer area escaped to the Dinka area and the Nuer who were living in the Dinka area escaped to join Nuer people. After that, it was raid after raid, killing after killing. That was the beginning of the destruction we are in today. Nuer [soldiers] would come at night and kill people, including women and children, and steal cattle. And, similarly, Dinka soldiers would go to the Nuer area and kill women and children and loot cattle. When Riek and Garang split, instead of fighting our common enemy, the soldiers came from Riek's side to ours … Our problem with the Nuer has never been a dead animal or grazing or anything. Our problem has always been the cow between us. The Nuer are a people who like cows a lot. Once they see a cow of ours, they take it. This has been our traditional quarrel with the Nuer. But we quarrelled only over the cow. This recent quarrel is not our traditional quarrel. When we used to fight, we used to fight with spears. This quarrel is a fight between two governments and that is why people are now killing themselves with guns.

By characterizing this conflict as 'a fight between two governments', the speaker was drawing on a conceptual distinction that had emerged during the mid-1980s between what people termed 'government wars' as opposed to 'homeland wars'. 'Homeland wars' grouped together all civilian-based conflicts, whether waged within or between Nuer and Dinka communities. 'Government wars', in contrast, encompassed the north/south conflict as well as military struggles between the SPLA and Anyanya-2 forces during the early years of this war (see Johnson, this volume).[8] These latter struggles began in 1983, when John Garang first clawed his way to the top of the southern military hierarchy by eliminating several Nuer and Dinka military leaders who had already formed their own rebel movement, known as the Anyanya-2 (see Nyaba 1997; Hutchinson 2001; Johnson 2001). Scattered and hunted by Garang's SPLA forces,[9] many of the predominantly Nuer Anyanya-2 leaders eventually forged an alliance with the Khartoum government, from which they solicited arms to fight against Garang. Although the bulk of the Anyanya-2 army was subsequently incorporated into Garang's SPLA during the later 1980s, several outstanding sections, including hundreds of Bul Nuer troops commanded by Paulino Matiep Nhial, remained firmly allied with the Sudanese Army. But, as far as most Nuer and Dinka civilians were concerned, these SPLA/Anyanya-2 battles were entirely 'government' affairs, as the following remarks by a Dinka church leader in Lokichokkio make clear:

> When the first groups [of trained SPLA recruits] returned from [their] Ethiopia[n training grounds], they came in different divisions, bearing different colours and carrying guns. They called themselves Buffalo, Tiger, Crocodile and other [battalion] names and said they were coming to fight [the 'Arabs'] in the south. During the time of the English, people were taught order. So people were not interested in engaging in local fights. There were no fights between us, the Dinka and Nuer, at that time. So, when these men returned, organized into battalions and carrying guns, we all considered them to be the government! They were composed of a mixture of Dinka, Nuer and other nationalities. So they were respected as a government. And, as for what was happening between the SPLA and the Anyanya-2 at that time, we [the civilians] had nothing to do with it. We were not involved because we felt that it was an issue between governments ... So, too, the fight between the government of Sudan and SPLA forces was to us a fight between two governments.

But, following the breakdown of southern military unity and the escalation of inter-ethnic violence after 1991, many Nuer and Dinka civilians began to identify their new-found sufferings with the birth of a new breed of warfare, which some

---

8. Linguistically, the term *kume* or 'government' in Nuer derives from the Arabic word for government: *hukuuma*. The term *turuk* in Nuer may also be used to mean both 'a government official' and 'an educated person'. This term dates back to the era of Ottoman domination during the nineteenth century. In Dinka, the term *turuk* may be used interchangeably with the term *akume* to signify a member of the educated elite as well as a government official. In Nuer, 'a person of the government/educated elite' would be referred to as *raan kume*.

9. It should be noted that the Anyanya-2 leadership initially included some prominent Dinka, who were also killed by Garang's forces. Many of these killings were carried out by SPLA Commanders William Nyuon Beny (a Nuer) and Kerubino Kuanyin Bol (a Dinka) during the early years of the wars. Some of the SPLA commanders who carried out these murders later defected from the SPLA and attempted to blame Garang for their previous atrocities.

called 'the war of the [southern] educated [elite]'. Local ethical codes of warfare that had been respected by generations of Nuer and Dinka combatants were completely ignored. Powerful AK-47 rifles and bazookas replaced the individually crafted spears and shields that had dominated Nuer/Dinka confrontations before that time (Hutchinson 1996, 1998). Brutally stripped of their former immunity from attack, women, children and the elderly were thrown onto the front lines. It became a 'coward's war', in which men, hiding behind their guns, targeted the weak and vulnerable (Jok and Hutchinson 1999).

Consequently, one of the questions the chiefs were struggling to answer at Lokichokkio was: 'Why did this terrible thing happen during the reign of the educated?' As one attending Nuer chief exclaimed:

> They used to tell us that the Nuer and Dinka fought each other because we are ignorant. We don't know anything because we are not educated. But now look at all this killing! This war between the Nuer and Dinka is much worse than anything we experienced in the past and it is the war of the educated. It is not our war at all!

Unlike the small-scale cattle raids that had periodically marred Nuer/Dinka relations in the past, this new 'war of the educated' appeared 'endless'. Whereas 'homeland wars' rarely lasted more than a few days before the chiefs concerned stepped in to quell the violence, the 'war of the educated' had proved impervious to countless mediation attempts. And, unlike the north/south 'government war', this new conflict appeared to have no overarching political objective. Rather, this conflict, the chiefs agreed, had been forced upon them 'from above'.

Although profoundly aware that the Khartoum government was manipulating southern military rivalries for its own purposes, many Dinka and Nuer chiefs in attendance at the Lokichokkio Workshop placed primary responsibility for the conflict squarely on the shoulders of the 'two doctors'. Several Nuer chiefs expressed bafflement and frustration at Machar's continuing efforts to mobilize Nuer recruits to fight against Garang's SPLA-Mainstream forces in Equatoria instead of sending them 'to chase the Arabs out of the south'. Several Dinka chiefs also expressed consternation at the steadfast refusal of Garang to make any conciliatory gestures towards the Nuer, even though everyone knew that the situation was ripe for a major SPLA peace initiative.

For these reasons, the chiefs concurred that the 'war of the educated' was not 'the usual quarrel between the Nuer and the Dinka'. As one Bor Dinka pastor present in Lokichokkio put it:

> This conflict has political elements. The old wars between the Nuer and the Dinka were settled quickly and quickly compensated [with the payment of blood-wealth cattle]. There was also intermarriage at the same time as the fighting. Our last fight [in the Bor area] was in 1976 and everyone was compensated. After that, we lived together as brother and sister. We travelled freely in Nuer areas … Even at the start of this conflict in 1991,

there were many Nuer living among us. Their cows were looted together with those of the [Bor] Dinka [during a devastating series of assaults by Nuer forces loyal to Riek Machar during late 1991 known as the 'Bor Massacre']. Those Nuer, who were living with us, fought on the Dinka side. They did not go back to the Nuer side. Those Nuer stayed in our area until there was great famine. Then all of us went together to [the Nuer town of] Waat [in search of relief food]. But the Nuer who were with us stayed with us. That is why I say this conflict is not strictly between the Nuer and Dinka but that it has a political element. Before the SPLA split, there were no tribal conflicts. Many people think that this conflict is now between the Nuer and the Dinka. But, before Riek decided to attempt a coup, there were no fights between us!

Another prominent Dinka church leader present at Lokichokkio hinted at a more radical solution:

What is happening now is a political fight from the top, not from the grass-roots. The old people at home just consider it a fight between John Garang and Riek Machar. They are just fighting over leadership. They will be surprised some day when people come to kill them. They will say: 'Why are we being killed?' … As we say among the Dinka: 'Once one fish is rotten, the rest of the fish will also be rotten.'

Encouraging his non-literate chiefly counterparts to think for themselves and to reject any simplistic equations between advanced schooling and wisdom, he warned them against accepting whatever Garang or Riek says as 'the truth'.

It is we who are to tell the leaders what is to be done, not they who are going to tell us what is to be done. This is in the nature of things. You, the chiefs, you cannot give up and say, 'We are not educated.' No! God made you and you have lived all these years. That education is there because, without that education, you would not have survived all these years!

Another highly literate and urbanized Nuer pastor took this line of argumentation even further:

Maybe it is education itself that has blocked our generation from solving its own problems. In past generations, people had to keep all their ideas in their heads. Who are the wise people now? The [educated] townees or the [unschooled] villagers? We, the educated, don't listen to one another. We should go back to villages and talk to people. And, if we don't listen to them, let them curse us!

Again and again, the chiefs pointed out that the educated southern military leaders who started the conflict were not the ones suffering. Rather, it was their uneducated counterparts, both in the villages and in the lower rungs of the military, who were dying. 'Power struggles are universal,' declared one Nuer chief at Wunlit: '[But] they should not affect the people!' Addressing himself to the soldiers in attendance at that meeting, he continued: 'When I was small, I thought the educated people were wise and intelligent. But now it is you [the rank-and-file soldiers] who are uneducated and who were left here to kill each other while you, the educated, go to stay in Nairobi.' Another Dinka pastor attending the Lokichokkio conference succinctly summed up the disproportionate costs paid by uneducated civilians and military personnel with a well-known proverb: 'When two elephants fight, it is the grass that suffers.'

More importantly, speaker after speaker rejected Garang and Riek's 'right' to project their leadership struggles onto their respective ethnic groups. This point was expressed most forcefully by a Lou Nuer chief present in Lokichokkio:

> When Riek rebelled against Garang, he did not succeed in overthrowing him. And Garang did not succeed in crushing Riek. We were caught between these two forces, Riek and Garang, and both were killing us! … Riek Machar was given birth by the Nuer; he was not the one who gave birth to the Nuer! The same is true of Garang. Garang is the son of the Dinka; he is not the father of the Dinka. Now they have created this fight. None of Riek's or Garang's children are dying in the field and both of them are still young themselves. We, the old people, are the ones being finished. Where will we be getting more children?

Turning to the Sudanese and American NSCC officials present, he pleaded to have them 'bring these two people – Garang and Riek – in front of us [the chiefs]':

> They are our children. Garang and Riek should order their commanders to stop this war. If they do not do this, we will curse them! … I am appealing to you, the church people. Is there no one who can reconcile these two? Is there nobody who can tell us that Riek is wrong and ask him to step down, so that we can all follow Garang? Or, if Garang is wrong, you [should] say: 'You are wrong' and all the people will follow Riek. Is there no one who can say this? … We, the civilian leaders, and we, the villagers, our opinion has been ignored! Even if we were to say 'You, Garang, you are right; lead us!' our opinion would be ignored. Or, if we were to say 'You, Riek, are right; lead us!' our opinion would be ignored. But many of you people here are highly educated. Please, can you not tell one of them to step down so that we can all follow one leader?

This chief's desperate call for someone – anyone – to reconcile these two seemingly irreconcilable men was heartfelt and struck a chord with many other Nuer and Dinka chiefs attending that preliminary peace meeting. So, too, did the speaker's complaint about having been silenced and sidelined by a new generation of educated leaders, who had little use for the traditional knowledge and wisdom of their elders.

But most of all, this Nuer chief denied Machar and Garang's claims to speak for or otherwise represent the political interests of their respective ethnic groups. Rather, attending chiefs at both conferences perceived these sharpening ethnic animosities to be the result rather than the cause of the conflict.

Second Hypothesis: 'The Arabs are the ones engineering this conflict'

In their efforts to identify the underlying forces perpetuating the violence, several Dinka and Nuer chiefs and religious leaders present at Lokichokkio pointed to the political machinations of 'the Arabs' and other foreign powers intent on draining the oil wealth of the south. One Dinka church leader at that event proposed the following explanation:

> When the present war started, some foreigners were involved in the split. The 1991 split was engineered by foreign powers. They influenced Riek. We have been knocking our heads [together] because of wealth and leadership [struggles]. The Arabs send politicians to disrupt relations between the Dinka and Nuer ... Right now, the Arabs are preparing to clear out all the Dinka from Yirol, who are weak from hunger. We are being used! The Arabs are the ones engineering this conflict. The people who are benefiting are the Arabs. It is not our war [at all]!

Other Dinka and Nuer chiefs zeroed in on Machar's political naivety and short-sightedness in thinking anything positive could possibly come from an alliance with 'the enemy'. Once again, differences in educational advancement were cited as aggravating factors:

> Our enemy, the Arabs, has controlled our people due to our lack of education. We have not been educated to handle responsibility. Our people only became local inspectors and local teachers. So we feel cheated and deprived. That is why we fight one another. The Arabs called Riek and said: 'Take the gun and kill Garang and we will give you power!' We kill ourselves because the enemy is playing tricks. What will Riek do? If we, the Dinka, are killed, will he fight the Arabs in the end? Now the Arabs will sit back and enjoy cursing and mocking us. Now, I say, tell Riek: 'You have been deceived! You should come back to your people.'

A Dinka pastor from Melut also present at Lokichokkio underscored the government's efforts to drive international relief organizations out of the south as a preliminary step towards draining southern oil deposits as well as stepping up the forced recruitment of southern children into the Sudanese Army:

> In my area, there is no peace. The Arabs are exploiting our differences. It has always been the intention of the Arabs to drive out relief organizations. And that is what they have done. The reason why the government wants my area depopulated is for strategic reasons to drain the oil and to take it up without

people witnessing what they are doing. The few [southern] soldiers who are there have no ammunition. In our area, there are no schools. And, if we go to Melut for education, the Muslims convert our children to Islam. When our children reach seventh grade or fifteen years of age, they are taken immediately to the north, where they are forcibly conscripted into the Sudanese Army. There is no peace in our area.

Other speakers at Lokichokkio focused on the government's role in promoting political rivalries and military confrontations among its three most important southern allies at that time, Riek Machar, Paulino Matiep Nhial and Kerubino Kuanyin Bol. Riek and Kerubino had both signed the '1997 Khartoum Peace Agreement'. There was no need to obtain Matiep's signature, since his loyalty to the national government and Sudanese army had not wavered since the start of the war. With the signing of that agreement, Paulino Matiep and Kerubino Kuanyin were officially integrated into the South Sudan Defence Force (SSDF), nominally under Machar's command. These alliances, however, did not last long. The government proceeded, first, to build up Kerubino's forces as a political counterweight to those of Machar and, later, when Kerubino abruptly broke away to rejoin Garang, concentrated more exclusively on strengthening Matiep. Maintaining Matiep's allegiance was especially important to the government because his Bul Nuer forces were strategically located between the main oilfields in the Western Upper Nile and SPLA strongholds in the Bahr el-Ghazal. Calling attention to the self-destructive nature of these men's intensifying leadership struggles, one Dinka church leader present in Lokichokkio invoked a parable about four animals:

At one time, a lion, leopard, hyena and fox agreed to live together in one place. The lion, however, said that he did not like noise. He said he would fight anyone who made too much noise. The leopard said he would fight anyone who dared to look him straight in the eyes. The hyena said he always wanted to sleep in a certain spot. And the fox said that he was going off to visit his sister. During the night, the leopard lay down in the hyena's spot. The hyena woke him and looked the leopard straight in the eyes. They started fighting. The noise of their fighting woke the lion and he, too, became angry and started to fight. All three animals were fighting one another. By the following morning, the lion, the leopard and the hyena were all dead. The lion is Riek, the leopard is Matiep and the hyena is Kerubino. The Arab is the fox.

The fact that the government was actively fanning political rivalries among its southern allies, however, was not accepted by everyone. There was one Nuer chief present at the Lokichokkio meetings who rejected this interpretation of events. In his mind, the fact that the government's most loyal Nuer ally, Paulino Matiep, was actively mobilizing forces to fight against Riek Machar at that time enabled this chief to assert that Riek Machar was not really 'with the Arabs'. This outspoken Dok Nuer chief, who clearly supported Riek, reasoned as follows:

This fighting is not because of Riek Machar! We are destroying ourselves! Our destruction should not be blamed on the Arabs or the white people. We, the Nuer and Dinka, are destroying ourselves. I never saw an AK-47 rifle before Garang came. [All] these guns were brought by Garang, not by the Arabs. The voice that says this problem was caused by the Arabs is wrong! Over all these years [of warfare], I never saw an Arab in my land until this year [when a contingent of northern soldiers was first dispatched to Ler, Western Upper Nile, in April 1998]. It was only in April 1998 that I saw my first Arab gun. It was captured by Paulino Matiep. It was a short-nozzled, Chinese-made gun. That was the first time that I ever saw an Arab gun. This meeting, if it is true, must recognize that our situation is very disturbing. You can't say there is nothing of you in it! Or me either. I will not deny that there are Dinka cows with us. They are there now. And I also know that there are Nuer cattle there with you [the Dinka] and you cannot deny it! So let us leave behind our cattle and children who have died. First, let us achieve peace and then we will ask about those things … You, the pastors, you must ask Riek and Garang why they are killing us. We, the chiefs, can unite. And, if Riek and Garang really want to lead us, we should have a [court] case with the Nuer and Dinka chiefs on one side and Commanders Riek and Garang on the other. We, the chiefs, will ask them: 'Why are you involving us in your fighting?'

Statements such as this one reveal the difficulties some chiefs faced in attempting to transcend the 'segmentary logic' that had guided so many intra- and inter-ethnic conflicts and alliances in the past. They assumed, in other words, that the enemy of an enemy is an ally. And, for this reason, it was apparently difficult for this Nuer chief to recognize the ways that the northern government was playing Matiep and Machar off against each other. For the government's tactics in this regard grated against his understanding of the decentralized nature of political alliances more generally (see Hutchinson 2001).

Third Hypothesis: 'This is a personal struggle between Garang and Riek over leadership'

Nuer and Dinka chiefs at both conferences stressed the urgent need for Garang and Machar to restrain their personal ambitions so as to ensure everyone's collective survival. One Nuer chief pointedly remarked: 'If all of the Nuer and Dinka are killed, who are Garang and Riek going to be leaders of?' Condemning the tunnel vision of both Garang and Riek, one Dinka religious leader in Lokichokkio remarked:

The chair that Riek is sitting in now is not the Nuer chair. It is for the whole south. The chair that Garang is sitting in is not the chair for the Dinka. It is for the whole country. They will both leave someday and the chair will remain there as a public chair. And, if we all rush to snatch the chair, we will break it. And then, the chair will not help us … It is the chair of leadership that looks for the leader, not the leader who looks for the chair.

In attempting to identify the root motives behind these men's seemingly interminable power struggle, participants at both conferences invoked a number of vivid metaphors and parables. One Dinka leader likened their power struggle to 'two men fighting over the meat of a buffalo they have not [yet] killed!' Another characterized Garang's reluctance to reach out to Machar with a story about two brothers, one of whom departs on a journey, leaving the other at home to watch over the herd. While the first brother remains away, the second brother is 'happy' because 'he is free to enjoy all the milk of the herd'. 'Although he would never say so, in his heart of hearts, he does not wish for his brother's return because, at that point, he would no longer have all the milk to himself.'

This colourful image of Garang's motives was given added credence by a startling remark made at Wunlit by Nhial Deng Nhial, who was the reigning Governor of SPLA-controlled areas of the Bahr el-Ghazal Province at that time. Up until the Wunlit Peace Conference, the official SPLA line was that Garang would readily welcome back Riek Machar (and any other southern dissidents) into the SPLA's fold, provided that Machar publicly acknowledged the legitimacy of Garang's leadership and abandoned all calls for internal elections. This was a relatively 'safe' position for Garang to adopt since he knew it was unacceptable to Machar. By the time the Wunlit Peace Conference gained momentum in late February 1999, however, Machar was emitting increasingly explicit signals that he was thoroughly disillusioned with the '1997 Khartoum Peace Agreement' and was seeking a face-saving exit. There was thus hope, at least in some Nuer quarters, that Machar would accept Garang's preconditions and thereby pave the way for the reunification of the SPLA. Misplaced as this hope may have been at that time, it completely dissolved once Governor Nhial Deng Nhial revealed a new SPLA stance during his opening remarks at Wunlit. Playing down any need for a formal merger of Machar's forces with those of Garang, Governor Nhial Deng called, instead, for 'a unity of objectives'. 'What is important in the struggle', he asserted, 'is not necessarily a unity of factions, but a unity of objectives and policies.'

Garang's failure to endorse publicly the conference's aims was a nagging source of disappointment and concern to many attending Nuer and Dinka chiefs. Some speculated that Garang worried that the restoration of peaceful relations between the Western Nuer and Bahr el-Ghazal Dinka would undermine the political prominence of his Bor Dinka allies within the movement. Other local conspiracy theorists suggested that Garang feared that such a peace would trigger a wave of desertions among his Bahr el-Ghazal Dinka troops, who formed the bulk of his rank-and-file recruits. Significantly, Garang's home region of Bor, which was heavily devastated by Machar's forces during the first months of the conflict, was already enjoying the benefits of a separate truce agreement at that time – an agreement negotiated by the SPLA with neighbouring Nuer communities during 1994–95. Once the security of his home region was established, Garang's detractors claimed, he showed little apparent interest in extending this peace to Nuer and Dinka regions further west. Indeed, many participants of Lokichokkio and Wunlit Peace Conferences interpreted the failure of any Bor Dinka chiefs to attend either conference as a sign of Garang's continuing obstructionism. Garang's supporters argued, in contrast, that Garang's endorsement of the Wunlit Conference, whether made public or not, would have been necessary for the proceedings to take place at all.

The overwhelming support for the Wunlit peace process offered by Commander Salva Kiir tempered people's lingering concerns in these regards. As the SPLA commander responsible for the security of the proceedings and a Bahr el-Ghazal Dinka, Salva Kiir had witnessed first-hand the tremendous destruction and hardships generated in his home area over the previous eight years. Throughout this period, Dinka communities of the Bahr el-Ghazal had been fighting defensive wars on two fronts, one against pro-government Baqqara Arab raiders coming in force from the north and the other against Nuer forces led by Riek and Matiep pressing in from the south and east.

Be that as it may, attending Nuer and Dinka chiefs were also highly critical of Riek Machar's refusal to compromise his personal ambitions for the greater good of the south. Several Nuer chiefs strongly condemned Machar's decision to negotiate separate peace treaties with the government during 1996 and 1997. But these chiefs also stressed that Machar's political errors must not be projected onto the Nuer people as a whole. Rather, they pointed out the need for Garang and his Dinka allies, more generally, to acknowledge the fact that the Nuer were capable of producing strong and able political leaders. One eastern Nuer chief, present in Lokichokkio, argued:

> If you are a Dinka, don't say that a Nuer cannot be a leader. Because, if you say that, it is like what the Arabs said when the English left. They refused to let us [southerners] rule. So, when you say to Nuer that they cannot be leaders, that is what we don't like, that is what we reject! If you are a Dinka or a Shilluk or an Anyuak, you can become a chief [leader]! But that is what the Arabs sought to deny. That was the tricky plan the Arabs devised. And is it not that [attitude] that brought this whole war on in the first place?

However, the political compromise Machar had reached with the National Islamic Front government, as well as the continuing predations of Matiep's pro-government militia, still left Nuer participants at these meetings wide open to charges that 'the Nuer are with the Arabs'. In the poignant words of one senior Dinka chief present in Lokichokkio:

> Now, you see, we [the Dinka and Nuer] are one people. The only differences between us are the [scarification] marks on our foreheads. I am asking you, my brothers: If you leave [us] and side with the enemy and you finish off our brothers in Bahr el-Ghazal, will you be living in the south by yourselves? If God helps me to defeat the Arabs, will I turn and chase you out with the Arabs? Or will we leave you there? Even if you run away with the Arabs, I'll tell you: 'No! Brother, come back.' It is my body that is small but my heart is as big as this [two-storey] building [we are meeting in]!

This charge of treachery elicited much defensive posturing on the part of some Nuer participants. One Nuer chief at Lokichokkio, for example, countered by saying:

What I know is that we [Nuer] have not allied ourselves with the Arabs! We thought we were going to be given an interim agreement for three [more] years. And, if the Arabs violate the agreement, then we will go back to war. Let us agree at this conference how we are going to chase the Arabs away from the south. I don't want to wait until tomorrow! ... We have fought with them for a very long time without success. So what I want from you is the best way to remove the enemy from our lands. What we did was not to ally ourselves with the Arabs. We said that, if the first tactic [fighting] has failed, let us try another strategy. We adopted another way of fighting with our enemies and not surrendering to our enemies ... We were told from the beginning [of the conference] that we should not keep anything hidden [but should air all our grievances] ... It is unfortunate that some educated Dinka are not here to hear what I have to say. What I don't like is when people get up and say that 'the Nuer are with the Arabs'. This makes me uneasy because I know that we are not with the Arabs. I always thought that it would be those educated people who would bring us together but sometimes I feel as though they do not want peace and reconciliation to be achieved. If we want peace and know that the Nuer and Dinka are sons of the same mother, then we should come together and reconcile and leave that [accusation] behind for another day.

Yet there were elements of ambivalence running through many of these defensive commentaries, as well as through more partisan endorsements of Garang and Riek's political strategies and leadership abilities. Garang was revered by many Dinka chiefs, not so much for the specific policies or military strategies he had adopted, but for the fact that his opposition to the central government had never been compromised. Unlike Riek Machar, Garang had come to symbolize continuity and determination in the south's struggles against northern oppression and economic exploitation. And thus many of his supporters feared that, were Garang to go, so, too, would the hope of a future southern military victory. Similarly, various Nuer chiefs present at Lokichokkio clung to the hope that Riek's decision to 'adopt another way of fighting the enemy' had not betrayed the just cause at the heart of southern aspirations of political equality and independence. Nevertheless, their attempts to counter Dinka accusations that Riek had betrayed the movement appeared forced at times. For example, the same Nuer chief who earlier argued that 'the Nuer are not with the Arabs' later reconsidered his remarks and added:

[On second thought] you, the Dinka, are right when you say that we [the Nuer] are with the Arabs. Way back, when we were all with Garang, was there not a [Nuer] clan – the Anyanya-2 – that was with the Arabs? You are lucky because all the Dinka were in one camp and we had some people with the Arabs. [At that time] Matiep was in Mayom [Western Upper Nile] and fighting on the side of the Arabs. The commander of Garang who was fighting against Matiep [at that time] was Riek! There were a few Dinka [with Riek at that time] but the rest [of Riek's troops] were Nuer. It was we, the Nuer of Bentiu, who were fighting against Matiep! … When Riek split up the leadership of Garang, Matiep came and joined together [with us and Riek]. What happened was Matiep came and we stayed together. But Matiep [soon] realized that Riek was not [really] with the Arabs. So, Matiep went back to the Arabs and he is now fighting us. If we were with the Arabs, what would be the reason [for us] to be fighting with Matiep now? Brothers, don't call us Arabs. If we want peace and reconciliation, don't say that again! … This war is a war against the Arabs! What is happening now is a struggle for leadership among us as southerners … What is killing us is that the Dinka want Garang to be a leader for ever! What is killing us is that the Dinka are wholeheartedly backing Garang and the Nuer are wholeheartedly supporting Riek … But Riek and Garang are our children. People can't finish on account of these two people! We don't know who they will be, but there will be other people who will come later who can rule. We are not Arabs and the Dinka are not Arabs. We are southerners.

Less than three weeks after these words were uttered, Paulino Matiep's Bul Nuer forces, armed and supported by the Sudanese army, drove deep into the central Western Upper Nile, attacking and burning the once the bustling market centre of Ler, Machar's 'hometown'. This raid was followed by two more devastating assaults during July and August 1998, which resulted in wholesale destruction of the town, which had served as the transportation hub for all international relief supplies flights into the Western Upper Nile.

Many months later, when Machar finally sent a high-ranking Nuer SSDF delegation from Khartoum to investigate the condition of Ler, it was met by an angry and deeply demoralized Nuer civilian population, which demanded to know: 'What does peace mean to you?'[10] When a northern Arab military officer, who had accompanied the delegation, responded by saying that he, too, was having difficulty understanding why Matiep's forces attacked Ler, he was immediately challenged by Commander Peter Paar, a Nuer SSDF field commander who had participated in Ler's defence. Commander Paar reportedly countered with words to the effect: 'You can't fool us [with your denials of involvement]! We were trained by your hand and we know your methods.'

The 1998 destruction of Ler marked the beginning of a new and far more brutal phase of the government's long-standing military campaign to depopulate the rich 'oil zone' stretching from north of Bentiu, the provincial capital of the Western Upper Nile, to the White Nile port of Adok in the south-eastern corner of the province. Whatever hopes Western Nuer civilians may have held in the viability of the '1997 Khartoum Peace Agreement' and in the political savvy of Riek Machar at the time of the Lokichokkio Workshop crumbled after the late June 1998 attack on Ler. Once again, the complex military alliances underlying the state of affairs in the Western Upper Nile shifted abruptly. Apparently fed up with recurrent orders by

---

10. This information was obtained from the Nuer head of this investigative delegation.

Matiep to attack fellow Nuer, a large contingent of Matiep's Bul Nuer troops mutinied in August 1998 and joined forces with Garang's SPLA. Undeterred, the government of Sudan forged ahead with a 1.6 billion-dollar oil development scheme, which included the construction of a 1,600 km pipeline to carry southern reserves directly to newly built oil refineries and export stations in the distant north. And thus, by the time the Wunlit Conference took place in February-March 1999, criticisms of Riek Machar's leadership abilities by attending Nuer chiefs were more vocal and blunt. One Bul Nuer chief, for example, compared Machar's political predicament to that of the sacrificial bull, Mabior. 'This evil of the Arabs', he began, 'was left to us by the British!'

> Now, you, the Dinka, accuse us of collaborating with the Arabs. But, if we had gone with the Arabs, would we be here? When Riek signed with Omar [al Bashir, President of Sudan], Omar tied a rope around his neck, like [we did around] the bull, Mabior, of the other day, which tried to struggle against the rope. His [Riek's] legs have also been tied by Mohammed Othman al-Mirghani. Riek cannot untie that rope now. Riek and his army have not been integrated into the Sudanese army. Riek's army has surrounded Bentiu and Omar [read: the Sudanese government] cannot move out. As for Matiep Nhial, all Nuer have united [against him].

Another attending Nuer chief turned to his Dinka counterparts and said simply: 'If you see that your brother has fallen down a well, what do you do? Cover him over or help him to climb out?' By late February 1999, when the Wunlit Conference began, many Dinka and Nuer civilians complained alike of being 'exhausted by death'. What ordinary rural men and women desired above all else was for Garang and Machar to reconcile their personal animosities and political differences so as to protect 'the few of us [who are] still alive'.

Fourth Hypothesis: 'If the commanders were listening to Riek and Garang, this fight would already have stopped!'

Not everyone who attended these two conferences was convinced that the conflict was being perpetuated by Garang and Machar. In Lokichokkio, several chiefs defended these men, saying that they were not the ones who ordered all these military raids on Dinka and Nuer civilians. One Nuer chief reflected back to 1992 when a contingent of Garang's SPLA-Mainstream forces, commanded by a Nuer SPLA officer named William Nyuon Bany, entered his area shortly after Machar broke away. 'What is killing us today', he claimed, 'has nothing to do with Riek and Garang!'

> Way back, when it all started, there was one battle [that took place in our area] led by William Nyuon [who was then the second highest-ranking SPLA-Mainstream officer after Garang] ... It was Garang who told William to go to Ler. We thought this fight was the fight of a mature/older man – that it was a fight from Garang in truth! There was nothing that Garang's

forces did except military operations. We knew that he was fighting to defend his leadership. His men never killed any civilians. They never took anything until the day they were chased out of the area by Riek's forces. That is why all the Nuer are saying, like myself, that the killing of women and the looting of cattle are not from Garang. When Garang sent William Nyuon, he sent him to attack a specific military target. And thus we have always asked the Nuer soldiers [in our area today]: 'Why do you go to Dinka areas and kill women and children and loot cattle? Last time, when William Nyuon came here, he never killed women or looted cattle. So why do you do these things?' [Similarly], if you were to go now to Nuer areas of the Western Upper Nile and were to arrest all of [Machar's] soldiers, I do not think you would find one soldier who would say: 'On such and such a day, Riek ordered us to go to the Dinka to loot their cattle and kill their children.'

This opinion was also expressed by a prominent Dinka church leader present at Lokichokkio who said: 'Leave Garang where he is! Leave Machar where he is! There are people near your byres who are holding guns. These people who are near us, they go and steal cows and rape women and then say that it is Riek Machar or John Garang. No! We will control these people, not Riek or Garang.'

Chief Malwal Wun, the most senior of all Nuer chiefs, also rejected the idea that Garang and Machar were the principal orchestrators of the violence. In a pivotal speech delivered at Wunlit, he challenged everyone present by saying:

You, the Dinka, are you in control of your children today? After robbing our cows, have these boys given them to Garang? You, also, the Nuer, those cows that you looted from the Dinka, have you given them to Riek? You have spoken as if Garang and Riek destroyed everything. When you, the Dinka, loot [Nuer] cattle, does Garang go with you [on your raids]? And you, the Nuer, when you loot Dinka cattle, does Riek go with you?

Everyone, he argued, must take responsibility for reining in the violent predations of 'our children'. In his eyes, the real problem lay with a new generation of armed Dinka and Nuer youths who knew little other than war and who had lost all respect for their parents and their chiefs. In his words:

You, the Dinka Chiefs, you asked why our sons went to Bilpaam [the main rebel training camp in Ethiopia] at the start of this war. They went for all of us! All these [young men] were our children. They were not the children of Riek or Garang. You asked how our problems began. They began with the Anyanya-1 [who fought for the independence of the south against the northern government during the first civil war (1955–72)]. The Anyanya-1 returned [home when that war ended] in 1972. On their return, they brought their rifles. They robbed people here and there. These children of today observed these activities. Thereafter, they went to Bilpaam to acquire guns of their own. They … became the Anyanya-2. These are the people who caused our problems today. When the Anyanya-2 came back [from Ethiopia], they did not respect the chiefs, the courts or the system. And the law broke down. Now we [must] blame ourselves.

It was thus the process of militarization itself – and especially the uncontrolled spread of firearms – that was now threatening to destroy everyone, men and women, young and old. Other speakers at Wunlit concurred:

The conflict is not between Riek and Garang now. It is now between the communities or between the young people. When the fighting started there were young girls who were found and each man rushed in to get a girl for himself. I know of a case in which three men were fighting over the same girl. The commander came and said: 'Instead of killing ourselves over this girl, let's kill the girl.' So he shot her. Soldiers are all the same in their way of thinking … What I know is that once we, the Dinka and the Nuer, acquired firearms things became very dangerous. We must go back to our traditional ways … I am an old man who wants to do something good for our people. A country where everyone has a gun is dangerous. There are thieves who have guns and who gather to make raids. And then they say this is what Garang wants or what Riek wants. We must do something about these people!

One Nuer Chief, present at Wunlit, appealed directly to the soldiers:

You, the soldiers on both sides, you have killed yourselves [for nothing]! But, now that we have reached peace, please join us. We must join hands to defend the oil in our area. The oil is for the south. We have not said that it belongs to the Nuer. The oil belongs to the south. But, if we remain separate and do not join hands, then the Arabs will take the oil. It is our problem, all of us, the strong people in the south, the Dinka and the Nuer. What Kolang Ket [a famous Western Nuer prophet once] said in a song is: 'People do not know their problems.' You, the Dinka, if you allow the country to be destroyed for the sake of empty chairs, it would be a mistake. The power people are fighting for is empty!

One Bul Nuer chief, whose section had rebelled against Matiep in 1998 and joined Garang, was even more forceful in his condemnations of the mindless violence and unrestrained greed of the soldiers:

Soldiers are like snakes. When a snake comes every day to your house, there will come a day when it will bite you! Brothers, help us today, we should ask the soldiers whether they have ended the quarrels among themselves. People are now calling us names and saying that the Nuer and Dinka are really bad people. I ask you, when we used to fight with spears, did it ever happen that foreigners were called to come and help us solve our problems? ... I think that we were really powerless [to prevent this conflict]. But if we were strong, we could raise a case against the soldiers! If we decide today to recover all the cattle looted from the Nuer and Dinka, you will definitely not find them with ordinary citizens. You will find the looted cattle on both sides with the soldiers ... I know very well that, even if we sacrifice one hundred bulls to seal the [peace] accord, whatever the accord will do, the problem we will still face at the end of the day is with the soldiers. Will the soldiers return to fight each other? Will we be in a position to restrain them?

What I know very well is that our soldiers are really, really, confusing us. Last time Riek Machar split from Garang and now Matiep has split from Riek. I am confused, really, as to what they are up to. What is it that they want? [What I think is that] our people are deceiving us! What they are actually doing is looting our properties! Like when Matiep split from Riek, he looted us. And Riek also looted the other side. They are actually just looting us! ... Traditionally, when there was a lion that was eating our cattle, what we did was kill that lion! Let us join hands, all southerners who have come here, and ask Matiep, our brother, if he wants to go with us or with Omar [el Bashir]. If he is not one of us, we should not let him take our oil. And, if Riek and Garang are deceiving all of us into having this peace meeting and they themselves do not respect whatever agreement we come up with, then there are actions we can take. One option is that we ourselves can rebel against them! We can find ways of organizing ourselves. Even if it means destroying the whole southern lands, we can do it!

This was not an idle threat. Riek Machar's subsequent return to the Upper Nile (after his January 2000 resignation from the government) stimulated an explosion of Nuer infighting, effectively paralysing their contribution to the liberation of the south. This new wave of south-on-south fighting threatened to spill over into Dinka areas. The main beneficiary of all this violence, of course, has been the Khartoum government. Beginning in 1998, western Bul Nuer forces led by Paulino Matiep and supported by government bombers and northern government troops systematically cleared the central Western Upper Nile of Nuer civilians in order to expand oil prospecting on the part of a host of Canadian, Swedish, Chinese, Malaysian and other international petroleum companies (see Verney 1999; Amnesty International 2000; Gagnon and Ryle 2001; Rone 2003).

Fifth Hypothesis: 'This is a war in which men are the aggressors and women the victims'

Unfortunately, there were no Dinka or Nuer women present at the 1998 Lokichokkio Peace Workshop. However, several leading Nuer and Dinka spokeswomen were given an opportunity to speak at the 1999 Wunlit Peace Conference. In general, their remarks called attention to gender disparities of power at the heart of the conflict – a conflict in which gun-toting men were portrayed as the agents of aggression and unarmed women and children as the primary victims.

One leading spokeswoman, for example, challenged the overwhelmingly male assembly of Nuer and Dinka civilian and military leaders present with the following remarks:

> Why have our children been dying since 1956? Since creation, when has a man ever died with a child in his womb? I do not expect you to reply. You, men, do you experience the pain of childbirth? It is the woman who gives birth and it is she who knows the pain. So don't play with the lives of our children. Your [real] attitude is to kill our children [but this] must not be done with deception. You can say whatever you like. We [women] are listening quietly. But, if you fail to see our point of view, we shall make a coup and refuse to share our beds with you! I know you will laugh. First, you think that we [women] will not make a coup or [you think] that we are married with cows [and therefore cannot refuse your bed]. We are no longer women of the past! We are now educated and we can put into effect our threat!

A roar of male laughter filled the meeting hall. However, this woman's words were sufficiently incisive to provoke a series of rebuttals as the conference unfolded. Follow-up remarks by a Dinka church leader and NSCC organizer, for instance, stressed the fact that women, too, had benefited from the conflict.

> What the women said [is] that they are no longer ignorant women but are educated and modern. I was moved by and will respond to what the women said. But you also, ladies, [share in the responsibility for this conflict]. Your sons die because you tell them: 'You don't have cows with which to marry' and you also say [to your husbands] that 'children cannot be weaned [and sexual relations recommenced] because there are no cows for [providing] milk [to the children]'. But we must all join hands to solve the problem.

Another male speaker responded more angrily: 'You women have claimed that you do not steal cattle! The thieves of cattle, are they not married? Don't you women ask your men, 'Where do you bring those cows from?' You are the first to drink the milk!'

However, because the women were not given a second opportunity to respond, their unique perspective on the conflict was subsequently sidelined by the dominant discourse set by attending Nuer and Dinka men.

Conclusions

This chapter, written during August 2001, has outlined five interrelated perspectives on the root causes of the violence that marred Dinka and Nuer relations during the decade immediately following the collapse of SPLA unity in 1991. While the multiple perspectives offered at the Lokichokkio and Wunlit Peace Conferences amply revealed the profound frustration and bewilderment many Nuer and Dinka civilian leaders shared in struggling to stem the bloodshed, there were a few overriding themes that dominated their discussions at that time. First and foremost, most participants at these events vehemently rejected the idea that this conflict was generated by some deep-seated enmity between Nuer and Dinka. Rather, they recognized that contemporary patterns of ethnic animosity and affinity were the result, not the cause, of the violence. As one Nuer chief at Lokichokkio declared: 'There is nothing in the hearts of Nuer and Dinka to make this conflict. Something happened which is beyond our powers.' Whether ultimately grounded in leadership struggles within the southern educated elite or in intensifying competition for economic resources between rival southern military factions and their rural constituencies or in the political machinations of the Khartoum government, this conflict erupted when 'people were one'. Furthermore, everyone agreed that the present conflict was of a different nature from that of the 'traditional' quarrel between Nuer and Dinka over the cow. 'If this were our traditional quarrel,' one Nuer chief at Lokichokkio remarked, 'we, the chiefs, could solve it. But this is something other.' The astonishingly rapid post-1991 militarization of Dinka and Nuer ethnic identities has also contributed to a radical reconfiguration of power relations between men and women, young and old. Heavily brutalized and armed bands of Nuer and Dinka youth, whether operating under orders from their military superiors or raiding for cattle on their own, have severely tested the limits of chiefly power and challenged the authority of their elders more generally. At the same time, women and children have lost their former immunity from outside attack and have become more vulnerable to sexual and physical abuse at the hands of armed men within their own communities (Jok and Hutchinson 1999; Hutchinson and Jok 2002). These lamentable trends cut across both sets of communities and give added urgency to ending the fratricidal bloodshed among south Sudanese. Whether or not the civilian-led Nuer and Dinka peace process will ultimately succeed in restoring inter-ethnic peace, there can be no question that the initial steps taken by Nuer and Dinka chiefs and religious leaders at the Lokichokkio and Wunlit conferences represented a significant advance towards reining in the violence of one of Africa's longest and most lethal civil wars.

Postscript: January 2002

As bleak as conditions in Nuer regions of the oil-rich Upper Nile appeared between 1999–2001, the Wunlit Peace Agreement successfully withstood all attempts to undermine it, whether mounted by diverse groups of government-allied Nuer militias or by the Sudanese army. Even more remarkably, the spirit of inter-ethnic trust and cooperation generated by civilian and religious leaders at the Lokichokkio and Wunlit conferences eventually cleared the way for a major realignment of southern military forces as well. On 7 January 2002, Riek Machar and John Garang announced that they had overcome their differences and would be amalgamating their forces in a newly strengthened and reunited SPLA, under Garang's continuing command.

Chapter 3

The Experience of Violence and Pastoralist Identity in Southern Karamoja

Sandra Gray

Introduction

In this chapter, life histories of Karimojong women are used to explore the cultural contexts of armed cattle raiding among Karimojong agro-pastoralists in north-eastern Uganda. Women's recollections of the circumstances of vital events (child births and deaths) provide critical information on the demographic effects of decades of violence in this population. At the same time, they illuminate the origins of modern intra-tribal raiding in southern Karamoja. Historical analysis suggests that current hostilities between Bokora and Matheniko territorial divisions of the Karimojong arose from their diverging political and economic interests in the decades since the 1950s, culminating in an alliance between the Matheniko section and the neighbouring Turkana against the Bokora section. The personal remembrances of Karimojong women, codified in a local event calendar (see Appendix), point to a disparate experience of droughts, famines and raids after 1970. Differential experience of these events is linked to the serendipitous acquisition of automatic weapons by the Matheniko, to the role of different territorial sections as either aggressors or victims in the ensuing struggle to retain or restore their pastoralist livelihood, and to a process by which the use of armed violence became synonymous with the Karimojong's social identity as pastoralists.

Raiding, Identity and Social Change in Karamoja

In August 1998, a baseline study of child morbidity and mortality was initiated among Karimojong herders in Moroto District, Uganda, as the first phase of a long-term study of the effects of economic development on human adaptability in a pastoralist population. Results from the preliminary phase of the research show an increase in infant and child mortality in this population between 1960 and 1980 and a concurrent decrease in completed fertility of Karimojong women (Gray and Akol 2000). Mortality of children prior to age five was found to be high in comparison with that for other pastoralists in East Africa (Leslie et al. 1999) as well as other populations in Uganda (Barton and Wamai 1994). Furthermore, Karimojong children who were measured cross-sectionally in 1998 were both stunted and wasted in their early childhood years, whereas older children exhibited pronounced maturational delays (Wiebusch et al. 2001; Gray et al. 2004;). As measures of human adaptability, these baseline findings are indicative of extreme environmental stress and intense selection pressure operating in this population (Gray et al. 2004).

The proximal, physiological causes of child deaths in all decades were measles, malaria, gastrointestinal disease and acute respiratory infections (Sundal et al. 2001); decreasing fertility appears to be linked to high prevalence of secondary sterility and a high rate of pregnancy loss. The distal, social cause of both demographic trends, however, is intra-tribal cattle raiding, which has been the critical factor in crop failures, herd losses, food shortages and disease outbreaks, which have recurred with alarming regularity in Karamoja since the late 1960s. Raiding and associated violence have also disrupted Karimojong social structure and marriage practices, contributing to an increasing frequency of informal sexual unions and, I hypothesize, to an associated increase in the prevalence of sexually transmitted diseases. Widespread raiding also curtails usage of health care facilities in the district, with the effect that expansion of these services since 1980 appears to have had little impact (Devlin 1998; see also Barton and Wamai 1994).

Any effort to explicate population and health dynamics in Karamoja since 1950 therefore necessitates examination of the sustained violence that appears to have shaped them. In an article, Gray argued that widening effects of Karimojong cattle raiding in Karimojong society stem from its increased savagery and its singularly internecine aim. Both characteristics were assumed to signify a transformation of ecologically-based inter-societal violence of a form that has been practised for centuries by pastoralists inhabiting the semi-arid savannah of what are now northern Uganda and Kenya (N.Y Dyson-Hudson 1966; Lamphear 1976; R. Dyson-Hudson 1999). Gray attributed the metamorphosis of Karimojong raiding and its introversion to a complex political and ecological process that undermined fundamental adaptive responses of Karimojong herders to unpredictable environmental events. She situated the blame for the erosion of the pastoralist system in Karamoja squarely on the shoulders of colonial administrators, who implemented an infamous series of destructive policies to promote economic development in the district.

The enumeration of these initiatives, variants of which have been applied in other pastoralist zones throughout colonial Africa with similarly disastrous social and environmental consequences (Behnke and Scoones 1993; Fairhead and Leach 1996; Majok and Schwabe 1996; Swift 1996; Gray et al. 2002), is an all too familiar recitation: the introduction of commercial agricultural and ranching schemes (Barber 1968: 214–20; Fleay 1996); the creation of politically expedient colonial boundaries that constrained pastoralists' mobility (Barber 1968: 153); the designation of vast tracts of dry-season grazing land as wildlife preserves (Cisternino 1979); and overdevelopment of water resources (McCabe 1990). The cumulative effects of these policies were overgrazing and degradation of fragile grazing lands. Other manoeuvres by the colonial administration permanently destabilized political relations among neighbouring pastoralists. These included manipulation of long-standing inter-societal hostilities to subdue those groups, such as the Turkana, who most actively and militantly opposed colonial hegemony (Barber 1968; Lamphear 1992); the use of force to bring recalcitrant herders in line with Pax Britannica; and the introduction of European weaponry into an already tense pastoralist/colonial, pastoralist/pastoralist mix (Barber 1968: 91–105).

The fundamental mismatch between pastoralist and colonial perception and practice that is exemplified by these strategies undoubtedly contributed to the escalation of Karimojong violence in the last five decades. At the same time, an exclusive focus on external pressures may be inadequate, inasmuch as it ignores autochthonous structures and process. First, militancy is a long-standing theme in political relations among pastoralists in this region, and this intra-tribal aggression has frequently taken the form of large-scale regional conflicts, which, in form and savagery, look suspiciously like war (Thomas 1965; Barber 1968: 158; Lamphear 1976, 1998). Given their central geographical, historical and political location within this militant mix of clans and tribes, the recent escalation in raiding by the Karimojong may represent simply the most recent shift in local power structures and control of resources, rather than a structural transformation of the magnitude originally conceived by Gray (2000).

Secondly, militancy appears to be inextricably linked to maintaining pastoralist identity among the Karimojong – as among most pastoralists in the region – and this association also appears to have considerable time depth (Lamphear 1976). While it is true that access to livestock as well as to water and grazing resources depends to a large degree on cooperation among neighbouring groups, such collaborative associations are ephemeral because droughts and livestock epidemics continually strain territorial and political relations (Little et al. 1999; Gray 2000). The history of the region suggests that those groups who were unable or unwilling to respond in kind to political and environmental contingencies – whether cooperatively or aggressively – ran the risk of losing their herds, their livelihood and their identity within pastoralist society.

In this chapter, recent Karimojong history is explored in relation to these two premises: first, that modern Karimojong violence represents a logical offshoot of its structural antecedent, and secondly, that Karimojong militancy is critical to the preservation of a primary social identity as pastoralists. Both hypotheses assume furthermore that any structural modification of Karimojong violence and Karimojong society that may have occurred in recent decades derives as much from internal as from external pressures. Socio-political fragmentation of the Karimojong polity and reshaping of Karimojong identity ensued from a collision along the fault line of these two sets of opposing forces.

In contrast to my earlier political and ecological reconstruction of the history of Karimojong violence (Gray 2000), the present analysis utilizes personal testimonies of over 300 Karimojong women, who were asked to recall circumstances surrounding the births and deaths of their children, parents, siblings and spouses. Vital events were then situated chronologically in relation to a locally compiled event calendar (see Appendix). Inherently subjective, the women's accounts nonetheless illuminate a cultural tradition shaped by violence and are testimony to both the collective and the personal impact of that violence. Inasmuch as the emphasis and tone of women's narratives vary according to their affiliation with different territorial subdivisions of the Karimojong, they also provide us with insight into the origins of intra-tribal conflict. The overarching motifs of modern Karimojong experience are killing, famine, disease and displacement, but inter-sectional variations on these themes highlight the central role of violence in the divergence of Karimojong social and political identities.

The analysis begins with a brief overview of the political ecology of cattle raiding in Karamoja. A summary of the results of 1998 fieldwork follows, to provide a demographic context for women's recollections. Next, selections from their narratives are used to infer underlying themes of recent Karimojong experience. Finally, I examine how this experience has influenced Karimojong identity and the modern uses of violence in Karimojong society.

Background to the Violence in Karamoja

The Karimojong are members of the Ateker cluster, a nexus of seven Teso-Karimojong-(Nilotic)-speaking populations linked by shared linguistic, cultural and historical traditions (Lamphear 1992; see Map 3.1). According to Lamphear (1976), the Ateker cluster arose as a result of the coalescence into distinct political trajectories of a diverse group of ancestral Para-Nilotic clans, sometime after the seventeenth century. A rich oral tradition attests to continuous social and political interactions among these groups throughout their prehistory and history.

Whereas it is true that encounters between different groups of herders often resulted in the formation of peaceful alliances through assimilation and intermarriage, they were equally likely to involve reciprocal cattle raiding, which was a prominent feature of the precolonial socio-political landscape. A number of studies, utilizing ecological-functionalist and Darwinian frameworks, have ascribed the phenomenon of inter-tribal raiding to the harsh and changeable ecological context of East African pastoralism. In this view, raiding served important political and ecological functions by redistributing people and herds during droughts and disease outbreaks (Dyson-Hudson 1966; Barber 1968: 217; Ocan 1992; Little et al. 1999; Gray 2000; see also Ellis and Swift 1988). Although such explanations may deserve the criticism that they justify pastoralist raiding and killing (Fleisher 1998), they do nonetheless take into account the historical frequency of natural disasters and the continual reconfiguration of the political landscape as a result of pastoralists' territorial adjustments to environmental reversals. Indeed, the history of the Ateker appears to be essentially a history of such responses (Lamphear 1976).

During a series of protracted droughts in the last four decades of the twentieth century, armed cattle raiding escalated along the Karamoja/Turkana (Kenya) border. In response, government campaigns were launched repeatedly to disarm herders and to induce them to abandon nomadic herding in favour of sedentary agriculture or ranching. In Karamoja, these programmes had disastrous economic, environmental and political consequences (Cisternino 1979; Alnwick 1985; O'Connor 1988; Okudi 1992). After 1979, when automatic weapons became widely available to pastoralists, the Karimojong retaliated with savage cattle raids against neighbouring tribes and ambushes of government troops (Ocan 1992; GOU 1994a). Intra-tribal raiding also became common, pitting members of the Matheniko territorial section of the Karimojong against members of two other major sections, Bokora and Pian

Map 3.1 Karamoja and approximate locations of the homelands of the Karimojong-speaking cluster of the Eastern Nilotes (Ateker)

(see Appendix). Today, Karimojong raids and related violence present a formidable barrier to economic development in the district, to the restoration of its infrastructure and to its integration into the national political structure and regional economy (GOU 1994b).

Demographic Research in Karamoja, 1998–99: Preliminary Findings

Between August 1998 and March 1999, life histories were compiled for 305 Karimojong women in Matheniko and Bokora territorial sections. These were all of the adult women in four clusters of agricultural homesteads, two of which were in Matheniko and two in Bokora (insecurity in southern Moroto precluded sampling in Pian). The dry-season herding camps for the Matheniko communities were located to the north-east and south-east, along the stretch of Turkana border between southern Kotido District (Jie) and Pian. Cattle camps of the Bokora communities were to the west and north. At the time of the fieldwork, violent encounters were frequent between Bokora and Matheniko herders along their common shared border in north-central Moroto, between Matheniko and Tepeth in the north-east and between Matheniko and Pokot, Pian and Tepeth in the south.

On the basis of our preliminary estimates (Sundal et al. 2001), mortality before the age of five years was approximately 25 per cent for children born between 1980 and 1989 and 19 per cent for those born between 1990 and 1999 (data for the 1990s are right-censored and therefore incomplete). Mortality before the age of ten years was 30 per cent in the 1980–89 cohort and approximately 20 per cent in the 1990–99 cohort. Estimates of under-five mortality in the 1980s are higher than estimates for other pastoralists in the region in the same decade (Leslie et al. 1999), and they are also high in comparison with most other districts in Uganda in 1994 (Barton and Wamai 1994). There was marked inter-decade variability in child mortality in Karamoja from 1950 to 1999: between 1960 and 1979, child mortality before the age of five years peaked at 32 per cent (higher than Kitgum and Gulu Districts in 1994), whereas 40 per cent of children born in that same twenty-year period died before the age of ten years.

Our results look somewhat different when estimates are calculated by territorial section. Mortality before the age of five years in the uncensored decade between 1980 and 1989 is 23 per cent and 35 per cent in the Bokora and Matheniko samples, respectively. By the age of fifteen years, 29 per cent of Bokora children and 41 per cent of Matheniko children born between 1980 and 1989 had died. In fact, in every decade since 1950, Matheniko mortality prior to the age of fifteen years is higher than that experienced in Bokora, although estimates for both districts are uniformly high. The difference is marked for the period since 1980, when Bokora under-fifteen mortality drops dramatically from an all-decade high of 46 per cent (1970–79) to its lowest rate, 14 per cent, between 1990 and 1998. In contrast, Matheniko mortality prior to the age of fifteen is 47 per cent or higher in every decade between 1940 and 1979 (56 per cent between 1950 and 1959). It never drops below 30 per cent, even in the (truncated) 1990s. In the total sample of 1,457 live births from 1940 to 1999, 36 per cent of Bokara children failed to survive to the age of fifteen years, in comparison with 44 per cent of Matheniko children.

Whereas infant and child mortality increased steadily between 1950 and 1980, completed fertility of Karimojong women in the sample decreased, from approximately seven live births among post-reproductive women who were born before 1945 to six live births among younger cohorts (Gray and Akol 2000). We have not completed our fertility analysis, but fertility appears to be comparable in the two territorial sections and is lower in younger cohorts than estimates provided in Barton and Wamai (1994).

Women's Narratives as Local History

Inter-sectional variation in mortality patterns and decreasing fertility provide us with our first insight into the differential experience of violence within Karimojong society. Women's narratives bring the background to these patterns into sharper focus. In Matheniko, disease and famine took a considerable toll in child deaths, but 40 per cent of Matheniko women in the sample also reported the deaths of at least one older relative (mother, father, husband or sibling) as the direct result of a raid or an associated act of violence. So common were deaths by violence in this section that accounts of these events have the self-conscious, stylized quality of recitations. Women's recall of killings and raids seem more litany than narration, more liturgy than personalized experience. The four excerpts below are a sampling:

[My daughter's] first husband was killed by friendly fire en route to Turkana to sell sorghum. He was killed by one of the people carrying the sorghum. She had three children by that man and then he was killed. He was a boy. Then she married Lokedikedi, within a year of his death. Lokedikedi was killed by Tepeth, en route to Loputuput, in the same year. She had one child by that man. Then she married Asoka and she has one child by him.

My father died long ago, in the year of Duarakile, when the Karimojong mounted a mass raid on the Upe, who retaliated. He died in the raid. My mother died before the man, of diarrhoea. My husband, Nangiro, was killed by Upe when he was herding at Naduket. He was a young man … I had five children … three are dead. One son was a married man, with a wife from Upe, but he was killed by Upe [Pokot] in Pian.

My father committed suicide when his cattle were raided by the Dodoth. His second wife, Ayelel, also committed suicide as a young woman. She left three children. One brother, Emeyan Karee, was killed by the Jie in 1998: he had a wife and child.

My father went for a meeting in Nakapelimyen, in cars. Before the meeting started, they were ambushed and he was killed, by Iteso. The bodies were buried in Nawoi. His third wife, Nalim Celina, died before he did: she was shot in the kraal, at Namalu.

In contrast, the stories of Bokora women are less enumerations of deaths by raids than detailed accounts of the lethal after-effects of raiding: disease, famine and flight. There is also a strong sense among women from this section of having been incidental victims, caught in the crossfire of Matheniko assaults and government retaliations. 'We were running,' recalled one woman, explaining why she did not take a mortally sick child to a health centre. In these stories, there is greater attention to the larger context and explication, as in the following three examples:

My mother died in 1977–78 of *lobulbul* [the swelling of hunger]: even her face was swollen and then her whole body. Then just the legs were swollen and she died. She had no treatment, because this disease did not want modern medicine [they could not get local medicines]. My stepmother was killed, with four of her children, in a raid by the Bagisu, in Mbale, where they had fled to escape Akoro [the famine of 1980). Two other children by this woman were killed in a raid in Teso. My husband was a home guard who was reassigned as a policeman under Amin. Then he was a soldier of Tito Akello, and he was away from Karamoja from 1982 until after Museveni's takeover. Three [of six] children are dead. One child died during the famines in the 1980s, of swelling [*kwashiorkor*]: she was listless. Her hair became very brown ... [we] were in Namalu then, where Sagal [her husband] had been transferred; the child was treated at Namalu Dispensary and Tokora Health Unit but all in vain.

My first husband was Lolem; this was not an official marriage but another man brought cattle to marry me, so Lolem's brother paid the bride price and I stayed with him. He died in 1975 of cholera, during the famine: an ambulance collected him, but he died en route to Mathany [St Kizito's hospital]. Then I was inherited by Adome, the brother who paid the bride price ... three [of eight] children are dead. These were all born in the period between Akoro and Nakoko [1980–85] after we fled to Masinde, where we stayed for the duration of the famine. Two babies died there, of diarrhoea and fever, because there was no hospital in Masinde. The last child died of cholera after we came home.

My mother hanged herself after the deaths of her husband and daughter. She despaired: 'What's the use of living?' My father died as a very old man, I was still at the breast. My sister raised me, I thought this was my mother. There was a brother too, but he left the district when I was young and nothing has been heard of him since. I have had five children; four are dead. One girl was grown and died of hepatitis around 1996. One son was a businessman, he was killed by Teso, after the murder of Apaloris: he was dragged from a vehicle and slaughtered. Another girl died in the meningitis epidemic, she was about ten years old.

The Uses of Violence

Woven into the starkly eloquent narratives of these women are broad themes of Matheniko ferocity and of Bokora suffering, which bring into high relief the divergent experience of the territorial divisions of the Karimojong during the calamitous closing decades of the last century. Differential experience of violence most certainly underlies mortality differences in the two sections.

The Matheniko were the first of the territorial sections to mount raids against other Karimojong and the first to adopt AK-47s as the weapon of choice for raiding. According to the reports of elders, as early as 1972 they had independently entered into an alliance with the heavily armed Turkana, who had been raiding widely in Karamoja since the 1950s (Thomas 1965; Barber 1968; see Appendix); subsequently the two groups mounted joint raids against their neighbours, including the Bokora and Pian. On the eve of Idi Amin's ousting in late 1979 and in the middle of a prolonged drought, Matheniko raiders ransacked his abandoned armoury in Moroto. As a result of this assault, they acquired thousands of AK-47s (Wilson 1985), and the balance of power in the competition for rapidly dwindling livestock herds swung decisively in their favour. Since 1980, they have pursued a policy of exaggerated militancy. Today, they have a reputation for being the most savage and most primitive of the Karimojong: Matheniko were involved as either aggressors or victims in nearly every raid in the district during our 1998–99 field season. Apparently proud of their notoriety, Matheniko men were more likely to refer to themselves as 'warriors'.

The Matheniko also have the greatest cattle wealth, accumulated at the expense of other sections and at great cost to themselves in terms of human lives lost. The majority of the Matheniko women interviewed were official wives, that is they had been 'married with cattle', and, for many of these, the bride price was in excess of 100 head of cattle. Nevertheless, many of these women had been married more than once, because their husbands had died in raids. Matheniko women's life histories are complex, entailing multiple husbands, multiple fathers of their children and many deaths.

The Bokora experience has been qualitatively different. Droughts in the 1960s caused many Bokora to leave Karamoja to seek casual labour in Jie to the north, in Teso to the west or in Kenya. In the 1970s, they took flight again, to escape famine and raiding by Jie, Turkana and Matheniko. Between the famine of the mid-1970s and the great famine of 1980, they lost their remaining cattle to the Matheniko. Although many Bokora women lost relatives to raiding, many more of their children, husbands, siblings and parents died of disease and starvation (*lobulbul*). Women recalled that, in 1975 and 1976, Bokora people tried to sell their children to traders in Mathany market, because they were no longer able to feed them (see Appendix). Most of the younger women in the Bokora sample – those who reached adulthood after 1970 – had not been married with cattle. Many had formed temporary unions wherever they went in search of work and refuge, with men who were not Karimojong. Some lost track of these men, as well as of the children they bore with them.

The Matheniko strategy for recouping their cattle losses in the droughts of the 1970s by raiding other Karimojong sections spelled economic disaster for the Bokora and Pian, many of whom subsequently fled the district. Others, however, took advantage of emergency relief programmes that sprang up during the 1980 famine or of food-for-work and primary education programmes operated by Christian missionaries. Some were forcibly resettled on government agricultural schemes (Wilson 1985). By exploiting the resources offered by outsiders they were able to survive the years of drought, raiding and famine from 1975 through 1984. They ultimately capitalized on the most important of these assets, education, to manoeuvre themselves into positions of political and economic power vis-à-vis the National Resistance Movement of Yoweri Museveni. Nearly a decade after the Matheniko, they procured automatic weapons as well and regained access to the pastoralist economy (see Appendix). They have continued to keep one foot firmly grounded in national politics and the regional economy as well.

Raiding and Social Identity in Karimojong Society

Any attempt to elucidate relationships between the widespread adoption of armed raiding and Karimojong social identity must begin with two elemental obligations attached to membership in the Karimojong political community in the past: first, Karimojong did not raid the cattle of other Karimojong (who were identified as 'brothers'); and, secondly, Karimojong did not turn their spears (the weapon of raids) against their 'brothers' but only against 'enemies' (Dyson-Hudson 1966: 209; on similar directives in Kalenjin societies, see Anderson 1986). If the meaning of spears is generalized to denote the most lethal weapon available, then inter-sectional raiding with guns constitutes an infringement of both injunctions. On the surface, the emergence of intra-tribal raiding with guns in Karamoja does indeed appear to signify a fundamental structural transformation of Karimojong society.

In fact, this is far from the reality, because the tenuous political affiliation implied by the designation 'Karimojong' (or Turkana or Jie, for that matter) is essentially an ecological artefact. As suggested earlier in this analysis, Ateker 'tribes' historically were transient territorial associations of clan-based herding units whose membership varied continuously in response to the recurring droughts, famines and disease outbreaks. The bounded political entity we denote as Karimojong is a colonial construct that equates political and territorial configurations of Ateker clans at the time of contact with distinct ethnic identities. It was an unstable structure from the onset, only temporarily – and ultimately ineffectively – contained by the colonial administration and Pax Britannica. Moreover, political identification of one group of clans as 'Karimojong' appears to precede the arrival of the British in Karamoja only by some fifty years (Lamphear 1976), and at the turn of the twentieth century their territorial identity was still being contested by neighbouring Pokot (Barber 1968: 157–68).

Thus, to attribute the apparent socio-political metamorphosis of the Karimojong to the erosion of some mythical tribal unity as a result of a relentless external process is to misread Karamoja's history and to ignore the expediency that defines Ateker politics. Identity with a particular political tradition has always been superseded by economic and social interests as pastoralists (Sobania 1991; Broch-Due 1999): where survival depends on having herds, to have herds is the only meaningful gauge of social location. Social identity in Ateker society is a question of membership, at a given moment in time, in one of two discrete groups: those with herds and those without. Political identity is similarly dichotomous: those who have animals to steal and those who steal them.

Informed by these socio-political realities, the prime directives of Karimojong politics may be restated: guns (and raids) may be directed only against enemies, who, by definition, cannot be brothers. Framed thus, their sense is equivocal and the identity of enemies ambiguous. Linked as they are to changeable status in relation to cattle, questions of identification as enemy or brother are resolved most expeditiously by firing or not firing a gun (or by taking or not taking an animal), and political affiliation is defined with similar expediency by who is holding the weapon and where it is aimed.

I propose that a struggle to maintain a primary social identity as pastoralists lies at the heart of the contest in Karamoja. Acting swiftly and opportunistically, the Matheniko applied their sudden wealth in automatic weapons to the implementation of a sublimely rational strategy, derived logically from the political tenets examined above and emphasizing the preservation of their pastoralist livelihood and their identity as nomads and herders. As a direct result of this strategy, the Bokora became refugees, both literally and figuratively: territorially displaced, without homes or families, but also culturally displaced, without herds and consequently without personal or social identity. From the perspective of economic survival, tribal affiliation was largely irrelevant.

In the late twentieth century, disparate historical trajectories thus produced two distinct variations on being Karimojong, both centred on pastoralism but arising from the alternate experience and use of violence. One vector in this bivariate social space describes the ultra-militant pastoralism of the Matheniko, whereas the other describes a gradual re-entry of the Bokora into the pastoralist sphere, but preceded by a decade of loss and alienation. The animosity between the two should not be underestimated, for their relationship, in the Ateker political sense, is as enemies and their views of the political universe are antithetical. A fierce adherence to conservative values and overt, systematic use of blunt physical force and automatic weapons sustain the one, while the other is supported secondarily by opportunistic armed raiding, but primarily by education, modernization and a capacity to manipulate national political and economic agendas to further local interests.

Premature deaths of both Matheniko and Bokora elders, either directly or indirectly as a result of violence since 1979, may have furthered the alienation of Matheniko from Bokora and Pian as well as from the Ugandan state. No sanctioned transfer of generational authority has taken place among the Karimojong since that recorded by Dyson-Hudson in the mid-1950s (Dyson-Hudson 1966), and an attempted transfer in 1998 was opposed by several of the most esteemed Karimojong elders (Gray 2000). At the same time, young educated men within the junior age-set as well as uninitiated members of the educated elite hope to expedite the transfer, with or without the approval of the elders. Inasmuch as educated elite are predominantly Bokora or Pian, national (parliamentary) recognition of their legitimacy as the governing class in Karamoja will serve only to heighten Matheniko resistance to government interventions aimed at resolving conflict, disarming herders or promoting economic development (the attempted transfer in 1998 was in fact facilitated by a government agency that oversees implementation of government policy in Karamoja).

Of Guns

What of the notion that the adoption of modern firearms irreparably altered pastoralist society by transforming the cattle raid into a criminal act (predatory violence for profit), as has been proposed by Fleisher (1998, 1999) and Hendrickson et al. (1998)?[1] Without question, the use of automatic weapons has been institutionalized in contemporary pastoralist societies in northern East Africa (Abbink, Tornay, Masuda, Volume I; Hutchinson, this volume) and with dramatic effects on other aspects of society and culture (Ocan 1992; Fleisher 1998; Gray 2000; Mirzeler and Young 2000). The gun trade, however, has thrived in this region since the end of the nineteenth century (Barber 1968: 101, 107–10), but the Karimojong were slow in substituting firearms for spears until the AK-47 presented them with a clear political advantage. Even so, the strategic realignment of the Matheniko with the Turkana occurred before the widespread adoption of automatic weapons in the region, challenging the conventional view that AK-47s themselves were the transforming force in Karimojong politics. Bokora informants recall the mid-1970s as a period when the Matheniko had very few guns of any kind but were already raiding widely against other Karimojong.

On the surface, the popular assumption that automatic rifles make individual raids significantly more lethal and hence more effective is a somewhat more compelling argument for the central role of the AK-47 in the transformation of Karimojong society. In reality this hypothesis remains untested, and John Lamphear, in a recent communication, suggested that evidence of pastoralists' brutality is present in history and prehistory, if only we look more carefully at oral histories and events commemorated in local calendars. Thomas' (1965) first-hand account from the early 1960s bears vivid testimony to the savage consequences of Turkana raiding for their Dodoth victims, and the Jie appear to have left few survivors during their campaigns against the Karimojong and Dodoth at the turn of the twentieth century (Lamphear 1976).

Conclusions: Violence, Identity and the Persistence of Pastoralist Society in Karamoja

Thus the modern version of Karimojong raiding does not represent a significant departure from the institution in place prior to the colonial era in either its intent, which is to increase or recover cattle wealth to sustain pastoralist identity, or its method, which is to kill enemies, or its intensity. The weapons (acquired largely by serendipity) have changed, as has the political location of the enemy, but then changing alignments and luck define Ateker politics.

---

1. Anderson (1986) identifies the criminalization of raiding with colonial restrictions on cattle theft. See also Ocan (1992).

The perspective that heightened violence transformed the Karimojong should probably be replaced by one in which the Karimojong transformed violence by exploiting guns and politics in characteristically opportunistic fashion to maintain their pastoralist identities. The ferocity of the Karimojong, and of the Matheniko in particular, not only has facilitated the recovery of the herds but also has been an effective strategy for deterring economic interventions emphasizing destocking, permanent settlement of the Karimojong, intensive agricultural development and the eventual dismantling the pastoralist system. Marauding Karimojong are acutely aware that, to settled people downcountry, they are alternatively viewed as savages, cowboys and warriors, and they exploit this notoriety to ensure the continued failure of economic development and sedentarization. At the present time, their promotion of an elaborated identity as 'warrior herders' may have much more to do with their relations with people from downcountry than with any fundamental structure of Karimojong society.

The preceding argument should not be interpreted as yet another justification for pastoralist violence. From an unabashedly Darwinian perspective, the costs in human mortality, human health and human development – and the latter is absolutely essential for social survival in the modern context – of maintaining the Karimojong pastoralist system and pastoralist identity may outweigh the benefits. The point at which the savagery that sustains such a system becomes maladaptive is the critical issue. Dyson-Hudson (1966) anticipated this evolutionary dialectic. In documenting the pattern of escalating intertribal raiding in the 1950s, he observed:

> I have no wish to caricature [the Karimojong] as incessantly hostile or preyed upon; if only for the reason that it is counter to common experience for political organizations to be perpetually in a state of aggressive turmoil, and so not to be assumed without strong supporting evidence ... [my data], I think, establish quite without doubt that Karimojong are so committed to their policies as to push them frequently to the point at which they may take both the lives and the herds of others, and risk the loss of their own. Dyson-Hudson (1966: 247)

Our own analysis of the mortality and fertility of Karimojong pastoralists in 1998 underscores the demographic risks intrinsic to such a strategy.

Herein lies the terrible paradox of modern pastoralism in northern East Africa. Today, as in the past, cattle-keeping is critical to human survival in these semi-arid rangelands; reports comparing the resilience of farmers and pastoralists in northern Kenya during drought have shown this to be true (Ellis et al. 1987; Nathan et al. 1996; see also Sellen and Mace 1999). As long as this is the case, we may expect pastoralists to defend fiercely their pastoralist livelihood and identity. Given the lack of either national or international support for subsistence pastoralism in Uganda,[2]

---

2. Exceptions include a recent agro-pastoral initiative by the Lutheran World Federation Programme in Moroto (BOLI 1998) and a now defunct restocking programme undertaken by OXFAM in Kotido (northern Karamoja).

however, raiding is perhaps the only strategy remaining to the Karimojong to ensure their continued access to cattle in the face of inevitable disasters. At the same time, modern raiding entails dramatic human losses, both directly to raids and indirectly through the effects of raids on other institutions. Our preliminary estimates, even at their lowest, show child mortality in Karamoja to be unacceptably high by any accepted standard. Indices of selection aside (Crow 1958), mortality of this magnitude today is synonymous with poverty, illiteracy, social disorder, political disenfranchisement, lack of infrastructure and alienation from both the national and global economies. In the twenty-first century, these are the distinguishing features of pastoralist culture. They do not, however, define either an infinitely sustainable subsistence system or a tenable modern society.

Appendix 3.1: Local Event Calendar for Bokora and Matheniko, 1909–99

Compiled by Helen Alinga Akol (Moroto) and Sandra Gray (University of Kansas),[1] 1998–99

Date	Bokora	Matheniko
1909	*Semei Kakunkulu*.[2] Named for the British agent responsible for bringing the Karimojong under British administration. Not clear who this is, but most probably Capt. P.S.H. Tanner.[3]	Same.
1910	*Ekaru a Lokijuka*[2] 'The Pusher'. Apparently named for E.A. Turpin, who acted to enforce the colonial administration of the then DC, H.M. Tufnell. Turpin was instrumental in building colonial infrastructure by forced labour and heavy fines imposed, in cattle, for failure to contribute such labour.[3] This date is questionable.	Same.
1911	Karamoja is declared a closed district by Acting-Governor Frederick Jackson. Tufnell (*Tapuna*) arrives with detachment of police to clear district of remaining ivory poachers and traders.[3]	Same.
1912	H.M. Tufnell becomes DC. It is most likely that Turpin arrives sometime after this, rather than in 1910. Pian calendar and Pazzaglia disagree on this date. Tufnell initiates building of roads and	Same.

other infrastructure.[3]

Local informant says this is the
year the poll tax was introduced.

1914	***Ekaru angolebei***.[2,3] Women are conscripted to provide grass and labour for building of colonial administrative headquarters in Moroto. Several elders recall this time.	Same.
1915	Alternative date for ***Ekaru a Lokijuka***.[3] Turpin most probably arrived in Karamoja in this year.	Same.
1916	Turpin becomes DC.[3] He remains DC until 1919; sometime in this period, there is a revolt of Karimojong against his draconian methods, culminating in the battle at Kayepas.[3]	Same.
	Poll tax (*ocur*) of 1 cent per adult male most probably imposed in this period, since it was introduced by Turpin. The tax apparently doubled every three years.[3]	
1917	Lokosawa selected as adviser for Karamoja District.[3]	
1917–18	Karimojong are recruited as soldiers for the First World War.[4]	
1919	Captain J.R. Roberts (*Ewoyaren*) is appointed DC.[3]	Same.
	Alternative date: Lokosawa becomes county chief in Pian.[2] This seems early, as it appears that the chief system and indirect rule were introduced somewhat later.[3]	Same.
1920	Battle between Suk and Pian at Nasogolo.[2] Local informants report a peace meeting at Lokales in this year as well.	Same.

1921	Karamoja is removed from control of KAR in favour of local administration. This appears to be the year that county chiefs were first appointed. In Bokora (Kangole), this was Lomanat (1921–44).[3] (Lokosawa in Lokales and Acia in Nabilatuk).[3] Ashton Warner is appointed first civilian DC.[3]	Lokong (1921–23) appointed county chief in Matheniko (Moroto).[3]
1922	Poll tax increased to 2 cents. This suggests the tax was first imposed in 1919.[3] Pian calendar says 2/-, but this seems unlikely.	Same.
1923	G.M.H. Lamb is appointed DC. Poll tax increased to 3/-[2] (unlikely, from[3]). Local chief, Acia (and Lokosawa[3]), is killed by Pian, in protest against taxation.[2,3] Several Karimojong were subsequently implicated and hanged.[2]	Longole is appointed county chief.[3] He is succeeded by Eria Acuka (1923–35) in the same year.[3]
1924	DC Lamb, whose harsh methods probably contributed to the violence of 1923, is replaced by DC Preston (called *Acolimatar*; alternatively, *Lojokatau*).[3]	
1925	Nothing recorded	Same.
1926	***Ekaru ngolo awaria akolong.*** Solar eclipse at 7.00 a.m.[2] ***Ekaru a lomee.*** Goats died in large numbers.[2]	Same.
1927	***Emorimor.*** 'It never stopped raining.'[3] Alternative date, 1931.[2]	***Ekokoro.*** The first bicycle, with the large front wheel, was introduced in the district.[4]
1928	Captain Rogers (*Lopokot*) appointed DC.[3] ***Ekaru a lobil angitomei.*** Elephants die of starvation in large numbers throughout Karamoja. ***Ekaru ngolo a ngimongo.*** People in Bokora ate this wild fruit because there were drought and hunger.	Same.

1929	Alfred Buxton briefly sets up first Christian mission at Lotome.[3]	
1930	Revd W. Owen opens first mission school for Karimojong children at Lotome.[3]	
1931	Alternative date for **Emorimor**.[2]	
1932	Preston is DC again by this year.[3]	
1933	The first of the Verona missionaries, Frs Molinaro and Vignato, arrive in Kangole (February).[3] Possible date of first populations census (Pian).[2]	Verona missionaries reach Moroto.[3]
1934	Nothing recorded.	Same.
1935	Approximate date of first *akiwor* of *Ngakomomwa* (women's age-set).	Elemudowa serves briefly as county chief in Matheniko. He is succeeded in the same year by Lokunoi (1935–44).[3]
1936	Nothing (first shop built in Pian[2]).	
1937	Nothing (Karamojong attacked Sebei[2]).	
1938	Nothing (elephant killed Nangiro[2]).	
1939	**Lokwakoit** 'white bones'. Drought or disease killed cattle in great numbers.	
1940	Nothing (cattles sales began in Pian, under Smith[2]).	
1941	**Ekaru ngolo abokere ngataparin**.[2,4] Construction of dams begins, with hoes.	
1942	Not clear, from any source. There appear to have been raids by Turkana (in Pian?), as well as raids on Sabiny, in Sebei, by Karimojong.[2]	
1943	Appears to be the beginning of about two years of prolonged drought and hunger. There may have been famine relief in this year.[2, 4]	

1944	Drought continues. It appears that the second group of Ngakomomwa (*Ngalingan*) are initiated, beginning in this year.	Drought continues. Lotud is appointed county chief. Some informants report that there was an outbreak of the disease *maring* in this year, in which afflicted people inexplicably were possessed by other people in the communities.
1945	*Lorengalaga.* 'Marriage bands rusted' because there was no ghee to grease them.[2,4] The drought is exacerbated in this year, with widespread famine. Many people still living remember this terrible year.	Same.
	Yakobo Lowok (1945–55) is appointed county chief.	Loduk Etwala (1945–54) becomes county chief.
1946	*Ngaduruko.* In this year, it appears that the *Ngakomomwa akiwor* was completed, and the first initiations of the succeeding set began. This sequence occurred because of the prolonged drought. *Ngakomomwa* (anthills) are associated with dry periods; sunshine is needed for the ants to emerge, while *Ngaduruko* bring rain; therefore, their *akiwor* followed at about the same time that *Ngakomomwa* was completed.[4]	*Lotokonyen.* 'People just sat and stared at one another' because there was 'too much sun' and nothing to eat.[4] Also *Ngaduruko.*[4]
1947	Nothing (in Pian, the Suk dance called *Naleyo*[2]).	Same.
1948	Formal *akiwor* of Ngaduruko.[4] The last initiations occurred sometime between 1952 and 1953.[3]	Same.
1949	A second or alternative year for the epidemic of **Maring**.[2] We got no consistent information on this event from anyone in Bokora or Matheniko: it may be that those who recalled it had in fact experienced it because they were in Pian at the time.	Same.

1950–52	***Aporodekeng***. This was a white man who killed two bulls for the initiation of Ngaduruko sometime in this period. He also got lost in the bush at some time in this period, an event narrated by some women in Bokora.[4]	
1950–51	In Pian, there were cattle inoculations, which were resisted by the people.[2] We found no local memory of these events in either Bokora or Matheniko.	Same.
1952	Earliest date given for ***Lotira***, but in Pian.[2]	Same.
1953	In Pian, agent Lorika and the *ekapolon* Nangiro were murdered by Suk.[2] Not remembered by Bokora or Matheniko. Dry season, 1953–54, is the period designated as ***Lotira***, 'the sun refused to set' and there were 'no clouds'. This was a drought period.[4]	Also ***Lotira***.[4] ***Lokulit***, for a man who was beaten in this same year.[4]
1954	***Todupak***, 'Plant it [anyway].' The drought continued into the planting season, and the Bokora planted anyway, in the hope that rains would come eventually and something would survive.[4] ***Longeu***, 'too much food'. The drought finally ended, and this refers to the abundant harvest of 1954.[4]	***Emathe***, 'locusts'. The rains came in Matheniko, and they planted, but locusts came and ate the crop, and they had to plant again.[4] Also ***Longeu***.[4]
1955	***Ekaru aruruma ka Akoba***. Yakobo is dismissed as county chief of Bokora, and many people remember that day.[3,4] Apparently, he had angered many people because he did not return Bokora cattle that had been raided by Jie, and yet taxed Bokora (cattle) heavily.[4] He is succeeded by Loyep (1955–59).[3]	Longora Lorupepe (1954–60) becomes county chief in Matheniko.[3]

1956	The *Ngigetei-Ngitukoi* relinquish power to the *Ngimoru-Ngikadokoi*, at Apule. This is the last recorded such event in Karamoja, and the last of the Ngimoru – the Ngibanga – are very old but still in power.[3,4]	Same.
	Some in Bokora report that this was the first year of the initiations of *Ngigetei* (those who replace their grandfathers).[4]	
1958	***Asapan a Ngigetei***. The better-known year of the initiation of the Ngigetei, in both Bokora and Matheniko.	Same.
		Toposa come to Moroto.
1959	Population census in Karamoja.[2]	Same.
	John Lolemunyang (1959–62) is appointed chief in Bokora.[3]	
1959–60	***Ekaru ka akan ka emeleku***. 'The hand and the hoe'. UPC and DP election campaigns.[2,4]	Same.
1960		Angyella Ibrahim (1960–63, 1964–76) is appointed chief in Matheniko for the first time.
1960–61	Dry season. ***Apanyanginyang***, a chief in Bokora, was killed. As punishment, Bokora cattle were confiscated by government, in helicopters.	Cattle were confiscated in helicopters.
1961	Planting season. ***Ekutelek***, 'caterpillars' ate the crop after planting, and people replanted.	
1961	Wet season. ***Lolibakipi***. 'Green water' (too much rain). There was so much rain during *erupe* that crops sprouted again before they could be harvested; nothing could be dried.[2,4]	Same.
1961–62	Late dry season. ***Erupe apa ongia***. The dry-season rains just continued into the next rainy season.[4]	Same.

1962	Weeding season. ***Namongo***. This was the widespread revitalization movement, which spread from the west into the whole district.	Same.
1962–63	October. *Uhuru.*	
1963	Dry season. ***Ekaru a mukuki.*** 'Spear'. Government soldiers raided homesteads and forcibly confiscated spears, punishing those found in possession of them.[4]	Yellow fever/hepatitis epidemic. Ngamilitia.
	Ekaru a nagilgil. 'Helicopters' seized cattled throughout Karamoja.[2,4] This may be the same event as reported earlier, in 1960–61.	Same.
	Yakobo Lowok (1963–73) returns as chief of Bokora.[3]	Timoth Loram (1963–64) becomes chief in Matheniko.[3]
1964	Nothing recorded.	Angyella (1964–76) is again appointed chief in Matheniko.[3]
1965	***Lomoroko***. The first comet.[4,5]	Same.
1965–66	Dry season. ***Logotho***, 'luggage'. There was hunger in Bokora, and people migrated to find food elsewhere.[4]	Ekaru Loziriao? Upe retaliation for a Mazeniko raid.
	Kalameriaputh. Bokora raided Jie at this place.[4]	
1966	Wet season. Mass raid in Dwarakile. Not sure if this is Bokora or Matheniko.[4]	
1966–67	Dry season. ***Dispersal of Ngibokora***.[4] In retaliation for Kalameriaputh, the Turkana and Jie raided Bokora in Lopei, causing them to flee Lopei for other areas throughout Bokora.	The year the helicopter crashed in Matheniko.[4] Year of many mosquitoes: yellow fever (Lonyang) and malaria outbreaks. Measles epidemic.
	People of Nyaikwae fled to Lorengacora in Iriri	***Kwari Kwar***
1967	Early wet season. ***Chepsekunya***. This place, in Sebei, was raided by Karimojong.[4]	Same.
	Lomoroko. The second comet.[4,5]	Same.

1968	***Lopetun***. 'Widespread'. A cattle epidemic in Bokora killed many animals. It began in this year.[4]	Angyella killed an elephant in Matheniko.[4] Ayopo was killed.
1969	The second year of ***Lopetun***,[4] in Bokora. Matany is under construction. The second population census.[2]	Same.
1970	Nothing reported.	Nothing reported.
1971	***Ekaru ka Amin***.[2,4] Amin seizes power. ***Ekaru alacia Amin ngitunga.*** 'Amin stripped people'. Traditional dress was outlawed and people were forcibly stripped; women were made to burn their skins, remove their wedding bands and crush their (then) glass beads into dust. Traditional hairstyles also forbidden.[4] ***Ekaru amunyarere ngitunga a Nawaikorot.*** Slaughter of Bokora at Nawaikorot. In protest against Amin's decrees against traditional dress, people dressed in their best ceremonial clothing and marched to Nawaikorot. There they were gunned down by Amin's troops and subsequently buried in a mass grave.[4] ***Kolera.*** Widespread epidemic. There was an immunization campaign and chloramphenicol capsules were distributed.[4]	Same.
1972	***Ekaru erisere ngimidi.*** Asians were expelled from Uganda. In Bokora, the two Asian traders in Kangole left (Babu and Lojulu). People remember them well. Mukuri raid in Pian?	In Matheniko (Moroto), the trader's name was Apa Lomagal.
1973	***Ekaru atwania akolong ecapio.*** 'The sun disappeared'. The second eclipse.[4,5]	Same.

Filippo Lokongo (1973–78)
becomes chief in Bokora.[3]

1973–75   By government decree, people were forced to grow cotton in Karamoja.

Same.

1974–75[4]   ***Lobulbul.*** 'Swelling'. Cattle disease and famine. In Mathany, people sold children to persons from other districts, because there was no food to feed them. People captured rats, which they roasted and sold for food. They also sold their beads and tools, anything that would get them a little food. The Bokora were dispersed widely across Uganda as a result of this famine: to Teso, Masindi, Iganga, Mbale, Busia, Kenya as well.

Matheniko began to raid Bokora cattle, using a few guns and pounding on noisy cans to make it seem that they had many guns. Bokora attribute the famine to the loss of their cattle to Matheniko raids.

'The people did not have salt'.

Ngamatidai, 'home-made guns'. Bokora began to manufacture home-made guns to fight Matheniko.

Nacoekale. Turkana raided Bokora cattle at this place.

Apalothiyel. This man was ADC-Bokora until 1976, real name, Rukua (from Arua). Rather than allowing Karimojong criminals to receive a trial, he simply slaughtered them outright.

Father Elia opened food-for-work farms in Lokopo.

1975   ***Amukad.*** 'Home guard'. The Matheniko raided the home guard kraal at Morulinga.[4]

Same.

Angyella finishes his second term.

1976–77   ***Ekaru ke ebuta.*** 'Sorghum didn't put out seed', although it flowered.[4]

1978   ***Elolimaata Ngiturkana ka Ngimatheniko Bokora*** [4] Raids against Bokora, by Turkana and Matheniko, intensify.

Same.

1979	***Ekaru eritarere Amin alo kicolong***. Amin is ousted, around Easter.[4]	Same.
		Tanzanians are in Moroto.
	Currency exchange, around September/October.[4]	
	Ekaru abwanguniata Ngimatheniko ngatomian. Matheniko get many guns after raiding army barracks in Moroto.[4]	
1979–80	Beginning of ***Akoro***, the Great Famine.[4]	Same.
1980	***Ekaru ngolo anyamere ngikolia***. The Great Famine, when people received fish as relief food.[4]	Same.
1981	***Emucele ka ekao***. 'Rice and peas'. The famine continues and people receive rice, peas and oil as relief food. Women do not conceive, and this is attributed to the amount of oil in the relief diet.	Same.
1982	January. ***Apaloris***. While on a trip to Teso to promote peace between Karimojong and their neighbours, Apaloris, a Matheniko elder, is killed by Teso. This event sets off a bloody sequence of events in the next two years,[4] beginning with Karimojong reprisal forays into Teso. Many people are displaced, trying to escape the violence and killing.	Same.
1982–84	Teso pull Bokora from vehicles and kill them on the spot in retaliation for Karimojong raids on Teso after Apaloris's murder.[4]	
1983	December. ***Ekaru amilica***. Militia from Acholi arrive in Mathany to confiscate cattle. They are killed by Bokora.[4]	Same.
	Bokora get guns (from militia).	
	Lokuuta's shop in Kangole is burned. This may have been 1984.[4]	

1984	**The burning of Mathany.** To put down the increasing violence in Karamoja and in reprisal for the slaughter of the militia, the government engages in a mission of destruction across Karamoja, beginning in Kangole early in the year and spreading eventually to Namalu, Mathany and Iriri. All of these towns were burned.[4]	Same.
1984	Warriors waylay vehicles at **Nakicumet**, robbing their occupants of everything, even their clothes.[4] Eventually, Atom, accused of being the ringleader of these bandits, is arrested and executed, after leading the troops directly to himself, in open challenge.[4]	**Nakoko.** 'Helicopters'. At the end of the year, helicopters bomb people and animals in Moroto district.[4]
1985	**Tito Okello** overthrows Obote. Some people remember this event.	
1986	**Ekaru a Mseveni.** Museveni comes to power.	Same. The kraals were in Dodoth.
1987	January. Currency exchange.[4] **Ekaru ka '3-piece'.** As part of efforts to disarm the Karimojong, Museveni's soldiers tie people's elbows and knees together and hang them from poles to force them to reveal where guns are hidden.[4]	Same.
1987–88	**Ekaru ngolo ka emogo.** 'Cassava'. There is hunger, and Bokora go to Teso to purchase dry cassava for food.[4]	Same.
1989	**Ekaru a tongetak.** Matheniko and Bokora resume full-scale hostilities.	
1991	The raid by Matheniko on **Alinga's kraal.**[4] Batanga is bombed by *oketa* (helicopter).	**Ekaru a lorionolup.** Nothing germinated. **Akoro** 'hunger'.
1992	Poor harvest. Relief food in Moroto. **Nangerr** coordinates distribution in Bokora.[4]	Same.

Ateregaegae. 'Meningitis'. There
is an epidemic in Moroto district.
Immunization campaign begins
in the middle of the year.[4]

Athuroi. 'Birds ate the sorghum'
in Bokora; there were many birds
that year.[4]

Natheperwae. A joint party of
Somali, Turkana and Matheniko
raiders raid this place in Bokora.[4]

Such a year is reported in
Matheniko also, but we are not
sure if this is the same year as in
Bokora.[4]

The Bokora pursue these raiders,
and manage to entrap them and
recover their cattle, after
slaughtering the raiders, at
Matakul, a desolate place where
'people drink urine' because there
is no water.[4]

1992–93   January. *Ekaru ka Apuno*.
Apuno, a Matheniko kraal leader,
is ambushed and killed by Bokora
at what is supposed to be a peace
meeting in Lotome.[4]

Same.

Turutuko. Bokora are killed here,
en route to kraals, by Matheniko.[4]

Dry season. *Ekaru ka Najie*.
There is hunger in Bokora, and
people go to Najie to buy sorghum.
Much migration this year.[4]

1993   The Pope visited Uganda in
February.

1993–94   Government troops bomb the
Bokora kraals at *Batanga*, in
Labwor? Lango?[4]

1994   *Akiriket a Athuguru*. Planning
ceremony in preparation for the
big Bokora *akiriket* at
Arengepuwa.[4]

The road to Apeitolim, through
Lokopo, is repaired.[4]

Ekaru a ngimomwa apa Mulelia.
This Turkana man told the
Matheniko to perform certain
rituals to ensure an abundant
harvest. They did so, and the
subsequent harvest was a good
one.[4]

Constituent Assembly elections.

(Wet season) **Ekutelek**. Army worms destroyed crops but the harvest was abundant **Erikaria Lokajekel ngino mira anamanat.**

1994–95   Dry season. **Ekaru a akiriket a Arengepuwa** A ceremony involving all Bokora clans, when they pray for the well-being of all Ngikumwae. This is the most important *akiriket* in Bokora.[4]

1996   Parliamentary and presidential elections.

Same.

1997   **Ekaru a Lookot**. 'Diarrhoea with blood'. There is a district-wide epidemic in this year of what appears to be dysentery.

Same.

1998   Cholera outbreak.

LC elections, March–April.

Asurui. Gupti was in Nadunget.

Akiriket a Nakadanya December. Appears not to have been carried out according to long-standing protocol, and is currently blamed for the escalating violence in Karamoja in early 1999.

Same.

Ekaru a gumere ngaauk a nakoit.

Lord's Resistance Army raid dispensary at Morulem. They are repelled and scattered by Karimojong warriors.

Ekaru Apalowau.

1999   Escalating violence between Matheniko and Bokora, and Tepeth and Matheniko.

January. Raid and massacre of Matheniko at kraals in Namalu, by Pokot and Tepeth.

February. Matheniko raid kraals at Narengemoru in Bokora.

Matheniko kraals at Apule raided by joint Bokora/Jie party.

March. Mseveni calls out UPDF troops to disarm Karimojong in surrounding districts (Kapcorwa, Katakwi, Soroti, Kitgum).

Land surveyor, Okac, killed at Camp Swahili in Moroto, by Matheniko criminals.

Ekaru ngolo alo Moru Ariwon.

No harvest.

2000–02	Measles epidemic.	Same.
2000	***Ekaru abwanguniata ngitunga alo Komolo***	Peace with Jie.
2001	Helicopter. Little harvest and drought.	Little harvest and drought.
	Lokopo Dispensary built.	Poor harvest. WFP distributed food.
	'***Operation***' (disarmament).	
2002	***Ekaru ngolo asurunio arianga ngitunga alo Kumana.*** The year soldiers dispersed the Karimojong in Teso and sent them back to Karamoja.	
	'***Operation***'.	
2003		Jie and Mazeniko herded together in Nangolapolon.
2004	***Ekaru ngolo atwanitor elap.*** Lunar eclipse, 5 May.	
2005		Famine. WFP distributed much food.
2006	The famine continued.	

Notes

1. For information concerning the compilation of the calendar, for copies, to supply additional information or to correct any events recorded, please contact Dr Sandra Gray, University of Kansas, Department of Anthropology, 622 Fraser Hall, Lawrence, KS 66045, USA. Additions or emendations would be especially appreciated, as we feel this is, at best, a very rough outline of local events.
2. Pian event calendar; preparer unknown.
3. Pazzaglia, A. (1982) *The Karimojong: Some Aspects.* Museum Combonianum n. 37, Bologna, Italy.
4. Local informants.
5. Turkana event calendar, compiled by Paul Leslie, Rada Dyson-Hudson and Eliud Lowoto.

Part II
Politics of Kinship and Marriage, Sudan and Northern Kenya

Chapter 4

Endogamy and Alliance in Northern Sudan

Janice Boddy

R esearch in colonial archives suggests that British officials in Anglo-Egyptian Sudan were deeply concerned that their subjects remain who they 'are': members of distinctive 'tribes'. Tribes were thought to be 'natural units' of administration, and 'detribalized' people a source of sedition and unrest. Moreover, a person's conspicuous traits – language, dress, skin colour, physique, ritual body modifications – were taken as markers of identity, indeed origins. Such views had an adventitious precedent further up the Nile that helps contextualize my discussion of sociality in nineteenth- and twentieth-century northern Sudan.

The argument rests on two obvious but sometimes neglected ethnographic points. First, academic anti-essentialism notwithstanding, those whose worlds anthropologists describe frequently anchor their relations in shared ontology. If assertions of belonging that inform self-worth and moral allegiance are held to reside in bodily substance or semblance, then the meanings of these must be explored. Secondly, genealogies are often social facts even when most suspect as a scientific tool. Problems arise from presuming that either 'essence' or ancestry predictably governs practice. Instead, I suggest, for those who live along the Nile in Arabic-speaking Sudan, practice informs ontology, and genealogies are largely prospective, devised more with the future than the past in mind. But, to indicate how and why this is the case, I must begin in the pre-colonial past, and move through the colonial period to the time of my fieldwork from the mid-1970s on.

Historical Context

Until the sixteenth century the area of which I speak – the Kabushiya region of the Nile some 200 km north of Khartoum – belonged to the Nubian kingdom of ʿAlwa, whose inhabitants were nominally Christian and matrilineal. Between the sixteenth and eighteenth centuries ʿAlwa was absorbed into the sultanate of Sinnar ('Sennar' in Abu-Manga, Schlee and the introduction to this volume), an indigenous semi-Islamic feudal state that formed around 1500 CE, whose nobility continued to follow principles of matrilineal descent. In Arabic documents of the time, commoners in Sinnar appear as clans and tribes, each occupying an assigned territory ruled by the sultan's appointee, often a sister's son (Spaulding 1982). The cohesion of such groups derived in part from kinship, but also from intimidation sanctioned by the state. For, as Kapteijns and Spaulding (1982: 30) observe, 'It was a fact of life that individuals who left the security of kin and community became masterless men, subject to robbery and enslavement.' Within these locales, land was worked communally, in ostensible usufruct from nobles or, increasingly during the eighteenth century, from Islamic merchant-divines, to whom the sultan granted lands unfettered by royal dues. The latter soon became enclaves where sanctuary was granted the distressed and maverick alike, Islamic rather than customary law applied and a nascent patrilineal bourgeoisie emerged around trade and the conversion of use rights into religiously sanctioned private property claims (Spaulding 1982, 1985; Bjørkelo 1989).

In 1821, Sinnar was conquered by Ottoman Egypt. The 'Turks' furthered economic differentiation by prescribing Islamic inheritance protocols and personal property rights throughout the Muslim north. They did so less from faith perhaps, than to compel the commodification of land, intensify agricultural production and generate crops for export. Accordingly, rights to shares in the annual yield of an irrigated tract (*saqiya*) were transformed into divisions of the land itself, which could now be registered individually and more readily bought and sold. This, along with the application of shariah conventions by Ottoman courts, caused a remarkable fragmentation of land along the Nile, where farming was limited to flood plains, basins and riverbanks. Spaulding (1982: 5) found that in pre-Ottoman Sinnar no harvest share under a twelfth of a *saqiya* was recognized, but, when shariah was strictly enforced, deeds implying the prospect of a seventy-two-thousand, eight-hundred-and-sixty-fourth-sized plot were not unknown. Holdings were often too small or scattered to be worth the costs of labour, water and other requirements to work them, to the benefit of those anxious to see land, in Bjørkelo's phrase, 'enter more smoothly the market sphere' (1989: 63).

Moreover, the Turks shifted the customary division *saqiya* returns to favour the landlord and, to draw their subjects more tightly into the commercial sphere, assessed taxes on farms and herds in coin or its equivalent value in slaves, desired as exports and soldiers in the Egyptian army. High taxes and the increased individualization of risk meant that less successful farmers were often forced to sell or indefinitely mortgage their land in order to pay.[1] Others simply abandoned their fields to appropriation by someone who redressed the debt (Spaulding 1982). Debt was also built up through the infamous *shayl* ('carrying') system, a form of crop mortgage entailing exorbitant rates of return; this too could lead to forfeiture. Those loath to relinquish their rights but unable to farm might arrange for their land to be held in trust (*amana*) and cultivated by others; even today sales of family land are highly stigmatized and such arrangements commonly made. Men with demographic luck and means could thus acquire workable parcels or rights to manage a consolidated estate – often at the expense of kin, especially women, who as daughters inherited one-half the share of sons.

---

1. In the area where I worked, there are two forms of land mortgage: *rahan*, legal under Islam, in which the owner makes over use rights in return for an interest-free loan. The loan, minus the owner's share of the crop, is repaid on an agreed future date and the land redeemed. The other is *damana*, in which the owner continues to borrow against his land over time. This may result in a debt so large it can never be repaid, whereupon the user can sue for a permanent transfer of rights. This, I was told, is forbidden under Islam as usurious.

To make matters worse, land-rich farmers preferred to cultivate with slaves than with paid help or tenants, since slaves had no legal claim to a portion of the yield. Free men unable to make a local living migrated out, some to the north, far more to the south, where they joined the ranks of petty traders and fuelled the ruin of insolvent fellow villagers by trafficking in slaves, the only 'commodity' that was not a vigorously enforced monopoly of the Ottoman state. The traders, for their part, had few options but to become complicit in the system that had dispossessed them and precluded their earning a living on the land. Though outmigration had long been practised in the unforgiving north, its escalation from the mid-nineteenth century on was untoward and firmly linked to human trade.[2] 'Migrants' wives stayed behind and, though little is known of their fate, it is likely that many took refuge with kin or, as widows, became second wives to propertied men.

Yet not all migrants became merchants. Given that Islamic principles allow women to own and bequeath land,[3] some men sought to salvage their fortunes by marrying beyond their natal hamlets, residing with their wives' families and cultivating their fields. In the late nineteenth century, famines and retributive marches under the Mahdi's temporal successor, the Khalifa Abdullahi, exacerbated dispersal, with many being killed or forced to flee. After the Anglo-Egyptian reconquest in 1898, the British encouraged resettlement and began to register agricultural lands; in the area north of Shendi where I worked, remnants of families found themselves distributed over the arable landscape, up and down the Nile.

The Problematic Focus on Tribe

In the early twentieth century, colonial officials sought to categorize, document and thus make intelligible the heterogeneous population they now ruled. But an obsession with origins made for endless difficulties in classifying riverain northerners, who were variously described as an amalgam or confederation of Arab and more or less Arabized Nubian tribes linked by a fanciful assertion of patrilineal descent from the Prophet's paternal uncle, Abbas. Collectively most were known as the Ja^caliyyin, though some non-Ja^cali claim Abbasid pedigrees too. The celebrated historian Harold MacMichael (who dominated the Civil Secretariat in the 1920s) argued that native genealogists wilfully erred by neglecting the stronger Nubian strains in their make-up (1922: vol. 1, 235). Such 'wilful' mistakes, he noted, made native genealogies unreliable guides to what had 'really' taken place and who people 'really' were. The discovery that genealogies are not statements of unalloyed fact was hardly surprising to MacMichael; he surely knew that this was not their point. But he seemed uninterested in what they were: records of present and potential social alignments.

In the 1920s, the formal adoption of indirect rule rested on the notion that 'tribes' could be reconstituted if no longer politically viable, or prevented from

---

2. See Spaulding (1982); Bjørkelo (1989: *passim*) on the Ja'ali diaspora, and ^cAli (1972: 11) on Mohammad ^cAli's monopolies.

3. Many women do not take possession of inherited land, leaving it in their brothers' care as insurance against divorce or widowhood. However, their children, especially their sons, are more likely to press their claims.

disintegrating if they were on point of losing their relevance. According to Sanderson and Sanderson (1981: 124), a circular from the Civil Secretary's office in 1924 spoke of the District Commissioner's task 'as that of "regenerating the tribal soul"'. Officials were convinced that authoritarian tribal structures had characterized even acephalous southern groups in the past and, with proper nurturing, their vestiges might 'develop and perhaps in time … give birth to genuine "chiefs"'. Thus, 'it became one of the major duties of DCs to discover by research this 'ancient governing organisation, and if possible to revive it'. As the colonial period drew to a close in 1953, the Civil Secretary's office crowed, 'The effect of these reforms was not only to restore but also to increase the prestige of tribalism.'[4] Tribes that had 'lost' their leadership and former 'coherence' were considered casualties of the Mahdiyya (in the north) or the slave trade (in the south), or both.

Granted, those events had been hugely disruptive; yet the idea that there had once existed a stable set of coherent, discrete, primordial social entities is difficult to defend (James 1977). More problematic still is the notion that 'tribes' so defined (rather than, say, family, village or herding group) had ever provided Sudanese with their most vital affiliations, leadership, moral direction and self-esteem.[5] Regardless, a host of challenges to the colonial government, from crime, to snags in the implementation of indirect rule, to the glimmerings of Sudanese nationalism, were put down to the calamity of 'detribalization'. Tribal institutions were held to provide a 'stable foundation' for governing; 'detribalization' was synonymous with trouble and strife (Collins 1983; Daly 1986).

A desire for precise classification remained strong to the end of the Anglo-Egyptian Condominium, dominated as it was by Britain. In preparation for its only census, undertaken in 1955–56 just as independence was declared, a circular entitled 'Tribes of the Sudan' codified people in descending order, by '"Race", Groups of Tribes, Sub-Group [of Tribes], Tribe, Sub-Tribe'. While conceding that such divisions 'are bound to cease having their original importance sooner or later', the document states it is nonetheless the government's aim to enumerate the population by 'race' and 'tribe'. Yet the list includes several queries. Should the contemporary Fung (Funj), for instance, be classed as 'Race I, Arab' or 'Race III, Negroid'? Should the Gaafra be included in 'Arab, Miscellaneous Central' and, if not, where?[6] Such preoccupations indicate why ethnographic research was encouraged by colonial authorities, whose requirements stimulated and helped shape 'the infant industry of anthropology', to use Robert Collins' phrase (1983: 165).

Yet ethnographic research was largely directed to the south, about which the British knew little in 1898. As for peoples of the north, because they were 'Arab' and Muslim and had fought either with or against the Anglo-Egyptian force, officials assumed they already knew a great deal about them. MacMichael, for instance,

---

4. 'Feature No. 253, The Position of Tribal Leaders in the Life of the Sudan.' Civil Secretary's Office, 16 July 1953. Sudan Archive, Durham University (SAD) 519/5/20 (Robertson).
5. See Robertson SAD 519/5/20–20b.
6. 'Tribes of the Sudan', n.d. (MacMichael) SAD 403/10/29–41.

reflected that 'if local customs and ways of thought were to be respected ... they must be understood. In the north this presented no great difficulty' (1954: 107). Arabic was the official language of government and a vernacular in the northern two-thirds of Sudan. Northern Arabic speakers who held low-level government jobs or were servants in officials' homes were the colony's most accessible subjects, deceptively familiar to British eyes. And they wielded a precarious power, being cultivated as collaborators (in the war against Mahdism) and appeased as potential zealots (believed amenable still to Mahdist appeal). It is tempting to suggest that it was their knownness, but also their purported state of 'decline' from a nobler Christian, Nubian age, and before that an astonishingly cultivated Meroitic (Pharaonic) past, that made history and archaeology the disciplines of choice for learning about them, rather than ethnography, which, according to MacMichael, was useful for fathoming inscrutable 'others' in the here and now. When ethnographic enquiry was undertaken among Sudanese Muslims before the Second World War, the preferred subjects were nomads, exotic and appealing to sedentary Europeans imbued with popular orientalist interests of the day.

Arab farmers and merchants – the *awlaad al-balad* (sons/people of the country) – were not just more familiar, they were also culturally akin to Egyptians, Britain's rivals for the hearts and minds of Sudanese under condominium rule. Neither noble enough nor savage, and eager for schooling when offered the chance, they were often disdained. Witness a young officer describing a northern road where 'bare-footed herdsmen with their flocks – and a sling or a spear – looking like Michelangelo's David, pass across in front of you under the telegraph wires, while every now and then a beastly Ford goes by full of young quasi-Effendis in tarbooshes – like people in Cairo or Port Said'.[7] To one former governor, the north 'was ... the most civilized province with so many of its sons in the educated class, and with tribes and tribal leaders long experienced in trade and travel up and down the Nile'.[8] They were knowable, and accordingly 'known'.

But for post-colonial ethnographers, too much had been assumed. Research in the village of 'Hofriyat' reveals, not least, that the self-professed patrilineality of the northern Sudanese is hardly the unitary form it was frequently taken to be. My informants say they are not readily separable into discrete ethnic categories, clans or tribes. All speak Arabic, profess Islam and typically deny connection to the Christian or Meroitic past (see also James and Johnson 1988). Local variants exist in customs at birth, weddings and circumcisions (funerals being largely scripted by Islam), and kin groups may distinguish themselves by loyalty to specific Muslim holy men or 'saints'. These concerns of daily life (*dunya*) are mainly the province of women, whose residential fixity is far greater than men's even now. Perhaps the most common assertion of identity one hears in the north is that of shared place of birth and residence when young. And, despite formal patrilineality, origin place may well be that of one's mother's mother, given the strong preference that women bear children

---

7. C.A.W. Lea to his parents, 2 February 1926, SAD 645/7/37.
8. H.B. Arber, 'Sudan Political Service 1928–1954', SAD 736/2/21.

in their mothers' homes and afterwards remain there for some time. Such claims must be understood in terms of the history of placedness and resource individuation described above. Yet place and kinship intersect, as we shall see.

Kinship and Endogamy

During the fourteenth century there began among riverain peoples a creeping shift of emphasis from matrilineal to patrilineal affiliation as Nubianness and Christianity gave way to spreading Arabness and Islam. I say 'shift of emphasis' because, owing to persistent endogamy, matrilineal and patrilineal connections are here intertwined, and relations by descent and marriage are convoluted, tangled and dense (as the simplified diagrams in this chapter suggest). Though sustained by ideology and naming conventions, patrilineality in northern Sudan must be understood as providing a line of bearing in a complex relational field that is inherently cognatic and supple. Marriages between 'close' (*gariib*) kin are not isolating, however; rather, they have significant potential to create both regional and group alliances. This apparently counter-intuitive claim (if alliance theorists be believed) depends on how endogamy is conceptualized and practised, especially in situations involving the multiple marriages of women and men (serial monogamy or polygyny). Alliances in endogamous northern Sudan are highly elastic; they may be intensified and 'enclosed' through successive close marriages, or abandoned or further extended as demographic, economic and political circumstances demand. They are by no means prescriptive. Identities available to be mobilized in this way are multiple and overlap, but ultimately refer to sets of embodied relations that endure in social memory and provide stable points of reference for several generations at least. Shared bodily substances passed to children through both their mothers and fathers both provide ontological grounds for kinship and alliance. Thus, when I would ask a person's clan or 'tribe', I was often told: '*nihna Saʿadab wa Busharab*' (we are Saʿadab and Busharab) or '*nihna Jawabra wa Mussalimab*' (we are Jawabra and Mussalimab), indicating either father's and mother's or (if the person's parents were patrilateral parallel cousins) father's and father's mother's groups.

If affiliation is considered by one's informants to depend on birth and legal parentage, then an investigation of marriage practices, not just rules and preferences, will go some way towards understanding how identities are produced. And in Hofriyat such practices refer to ontology. According to local conception theory, a child's bones and sinew (hard parts) are formed from the semen or 'seed' of its father, while its flesh and blood (soft parts) are formed from its mother's blood (see Holy 1991: 48). Unlike some farming peoples, who imagine the womb as passive – as soil that nurtures but adds no substance to the crop (Delaney 1991) – for northern Sudanese, maternal contributions are active, dynamic and contributory. A mother's body is the source of her child's bodily fluids and tissues that animate its 'masculine' bones, just as water ensures the growth of crops and staves off desiccation in the desert north. The staple food, *kisra*, made by mixing dura flour obtained through men's farm labour or wages with water that women fetch from the Nile or village wells, supplies a material metaphor for human progeny and the capacities they contain.

The merging of seed and blood makes all bodies composites of male and female substance. Kin participate in each other's bodies and persons: bodies are neither

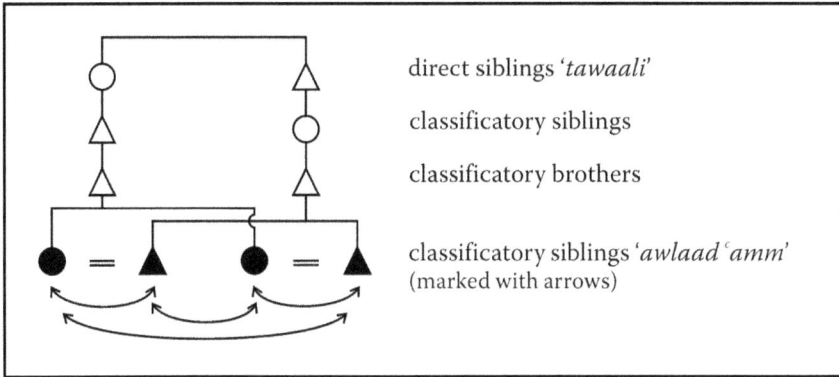

direct siblings *'tawaali'*

classificatory siblings

classificatory brothers

classificatory siblings *'awlaad 'amm'*
(marked with arrows)

Figure 4.1 Classificatory *awlaad 'amm*

unique nor finite in this sense, and a person is unthinkable except in reference to kin. One's social and material composition is wholly shared with full siblings of either sex, making sisters or brothers 'natural' substitutes in a marriage if one of them has died. Yet the mix of parental matter is bound to separate over the generations, for only sisters transmit maternal blood, only brothers transmit paternal seed. Moreover, a body is said to dry up as it matures: skin loses its moisture, womb blood no longer pools, flesh slackens and ultimately, with death, decays. Only the bones remain. Maternal blood is ephemeral, suggestive of all that is transient in the temporal world (Boddy 2002).[9] Paternal connection prevails, structuring bodies and social relations through time; indeed, the rigid appendages of the body – foot, calf, thigh – describe progressively inclusive generational levels in local patri-genealogies. Maternal relations are the tissue that binds paternal limbs into a social whole. When they fall away as memories fail, the 'bones' disengage. It is only then that lineages emerge distinct – unless, that is, connection is regenerated by successive marriages among maternal kin. Here lies a clue to the preference for close unions, why most first marriages are between first or second cousins of any derivation and virtually all are contracted within the village area. A consummate village is physically and socially integral, an enduring moral body repeatedly fortified and contained (Boddy 2002).[10]

However, northern Sudanese also use extensive kin classification, which supplements the transmission of bodily substance by creating further prospects for renewing past relations (Figure 4.1). Some implications of this were explored by the Kronenbergs (1965), who showed how, in the context of patrilineal endogamy, kin classification ensures the retrospective assimilation of uterine relations into the

9. See Holy (1991) on the Berti of Darfur. For a comparable case in Anatolia, see Delaney (1991). For consideration of how these issues play out in northern Sudanese kinship, see Kronenberg and Kronenberg (1965). On marriage patterns and the importance of maternal connection in historical Sinnar see Spaulding (1985).

10. This may well be related to the territorial 'enclosure' prescribed by the Sinnar Sultanate, and possibly the Nubian kingdom of 'Alwa before that. See Spaulding (1985).

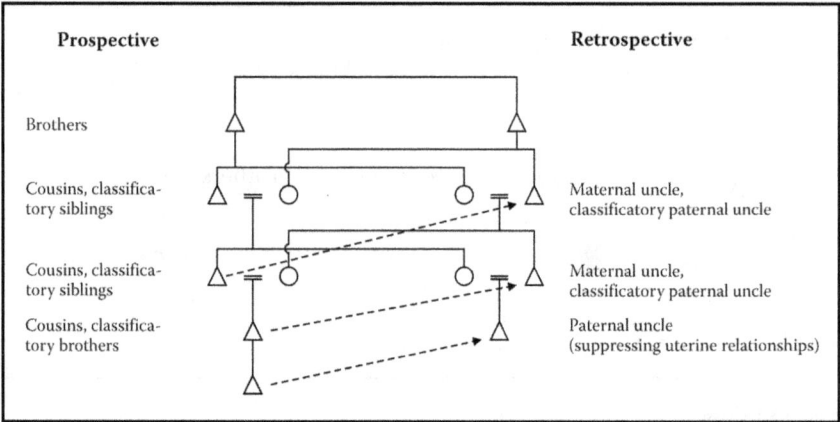

Figure 4.2 Constructing kin relations prospectively and retrospectively (adapted, with modifications, from Kronenberg and Kronenberg 1965: 241, Fig.5)

agnatic line: maternal kin become patrilineal when intervening female links are suppressed (Figure 4.2). No matter how convoluted actual relations may be, classification allows for the continuous idealization of patrilineages in social memory. It also allows a variety of close marriages to be considered exemplary, that is, as those between cousins whose fathers were 'brothers' (Figure 4.3).

Kin classification has broader implications still, for the children of siblings of either sex are considered 'siblings', if not immediately (*tawaali*) so. And in time this lack of immediacy may be forgotten or suppressed, to the point where 'cousins' move temporally forward, becoming socially closer, their ties more direct than they would otherwise be. Marriages between the children of actual or classificatory siblings of either sex are considered 'close' – *gariib* – and in the 1970s and 1980s accounted for some 70 per cent of all first marriages in the area where I worked. Most first marriages follow a centripetal, inward-looking social dynamic; second marriages, however, provide a centrifugal complement, for, unless people are fulfilling sororal or leviratic claims, they typically find spouses further afield when marrying a second time. Yet both close and relatively more distant marriages are subject to the consolidating tendencies of shared bodily substance over time.

This suggests that kinship in northern Sudan is not just cognatic but also prospective. For, in seeking mates for their children, people build on past relations or create them anew so as to generate future opportunities for alliance: gain access to new resources, strengthen political and economic support and widen the group with whom one shares moral dimensions of personhood. The sequential marriages of a woman to two or more unrelated or distantly related men are likely to create a cohort of uterine siblings who belong to different 'patrilineages', while the sequential or simultaneous marriages of a man to two or more unrelated or distantly related women may create a cohort of agnatic siblings whose mothers belong to different

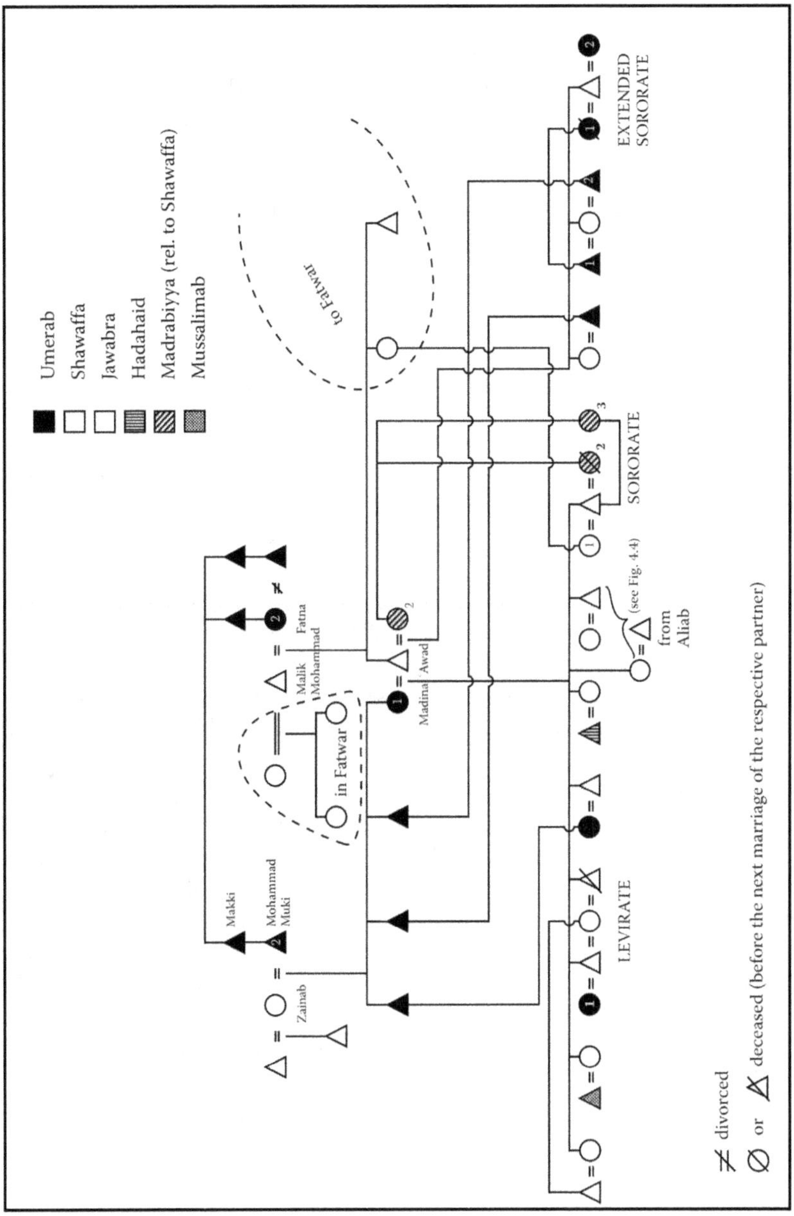

Figure 4.3 Grandchildren of Malik Mohammad and their marriages

lineages. In either case, should at least some of those siblings' children marry, they will marry 'close'. Some examples should help to clarify.[11]

Malik and His Kin

Sometime in the late 1870s or early 1880s, Malik came south to Hofriyat looking for land, a place to settle and descendants. He had a wife further north and several daughters but no sons. In Hofriyat he met Mohammad, who invited him to stay. Mohammad's people had not been in the village long, but his wife, Zaynab – a non-relative – belonged to a well-established family there. Mohammad was Zaynab's second husband; her first had been a cousin, with whom she had had a son. Mohammad cultivated his wife's family land, perhaps as a tenant. At the time, he was also caring for several daughters of his late father's brother (his *camm*). Because there were few men in his household, Mohammad may have been seeking husbands for his wards or labourers for his family's fields, or both. One of his wards was a cousin, Fatna, who was newly divorced from their mutual father's brother's son. The newcomer Malik saw Fatna *gacada sakit* (sitting silent/empty) and said, 'Maybe I can give you descendants, my sister.' She agreed to marry him because 'she'd had no luck with the men of her own tribe (*gabiilata*)'.

Malik and Fatna had a daughter and two sons (Figure 4.3, first descending generation). The daughter and one son married kin in Malik's natal village of Fatwar and went to live there. The other son, *cAwad*, remained in Hofriyat. Malik and Mohammad arranged that he marry Mohammad's daughter, thereby consolidating the bond between their fathers. *cAwad* later married again from among his Fatwar kin, and brought his second wife to the village. *cAwad* and his wives had six daughters and six sons (Figure 4.3, second descending generation). They in turn entered into a total of seventeen marriages, seven with members of Mohammad's lineage. The greatest number of these involved *cAwad's* children by his second, Fatwari wife, who required greater residential legitimacy in Hofriyat than his children by Madina, the local wife. Two other marriages built on links to Madina's mother's kin (through daughter or son exchange). Two more sons, but no daughters, married women of Malik's family in Fatwar whom they brought to Hofriyat. One of these men also married the sister(s) of his father's second wife after her death. Another daughter married a relative of Malik's who had settled in a nearby village, Aliab, suggesting that Malik may not have migrated south alone. The two remaining daughters married men belonging to other established families in Hofriyat. Links were thus consolidated in several key directions, creating a new endogamous cluster. Whereas Malik and *cAwad* had had to share-crop, their descendants soon acquired enough land from maternal and affinal kin for them not to have to do so.

Marriages of these grandchildren among themselves and with kin in the three putatively distinct patrilineages have continued over the generations. So has the exchange of personnel – women and men – between villages, depending on

---

11. All examples are drawn from field data, most of it previously unpublished. The following family history is based on conversations and formal interviews conducted between 1976 and 1984 with several descendants of the long deceased principals.

demographic conditions and economic resources available at any given time. Here alliances between places and people have been created and intensified by consolidating lateral and lineal relations through the marriages of close kin. It is not necessary that every member of a family wed members of only one other for an alliance to become useful and secure. But it does require that the association be reproduced by some. While such matches may seem lineally exogamous in the conventional sense, villagers deem them 'close'. They are considered, in other words, to be forms of 'in-marriage' or endogamy in that they defensively envelop and thicken existing ties.

Recall that Malik's first wife had remained in Fatwar with their daughters when he moved to Hofriyat. These daughters became place-holders for their father; so too for his Hofriyati offspring. Ideally, co-wives do not live together in the rural north but are housed in different villages. While distance helps to offset friction and maintain at least an illusion of the religiously prescribed equal treatment of co-wives, it also ensures that a man's children are territorially distributed and his connections wide. Given that children inherit from their mothers, this may mean that land and other resources become available to kin who were once 'outsiders' through marriages among the grandchildren of a polygynous man. At least three men whose ancestors had left Dongola decades before and settled in Hofriyat travelled to their ancestors' origin villages and found second wives among distant kin whose fathers had land but few sons or none. Their progeny have continued the alliances not only between Dongola and Hofriyat, but also with their satellite enclaves in Khartoum and Omdurman. The men thus gained descendants and opened up prospects for their expected grandchildren in areas distant from Hofriyat: access to commercial networks, foreign work permits, educational opportunities and government jobs, to say nothing of moral certainty, embodied trust.

Under certain conditions such manoeuvres may lead to conventional 'lineage endogamy', as the following case details.

ꜥUmar's Family

ꜥUmar is a great-great-grandson of Malik in the paternal line (Figure 4.4). His father's brother, Hamid, married in the nearby village of Aliab, where ꜥUmar's paternal grandfather had grown up. Hamid's non-kin wife, Mariam, was the only child of a couple with several hectares of land who needed a son-in-law to run their farm. Upon her father's death, Mariam inherited the bulk of the family estate. Soon she and Hamid sought a son-in-law for their daughter, Atiya, who stood to inherit a portion of Mariam's land. ꜥUmar was chosen – after all, he and Atiya were patrilateral parallel cousins – and the newly-weds moved into Atiya's parents' home. ꜥUmar and Hamid cooperatively farmed Mariam's land, and became successful producers of onions, an important regional crop. One of ꜥUmar's sisters, both of his brothers and Hamid's paternal half-brother's son, Wida'a – who was also ꜥUmar's mother's sister's son – later married Atiya's remaining siblings, a classic case of kin consolidation.

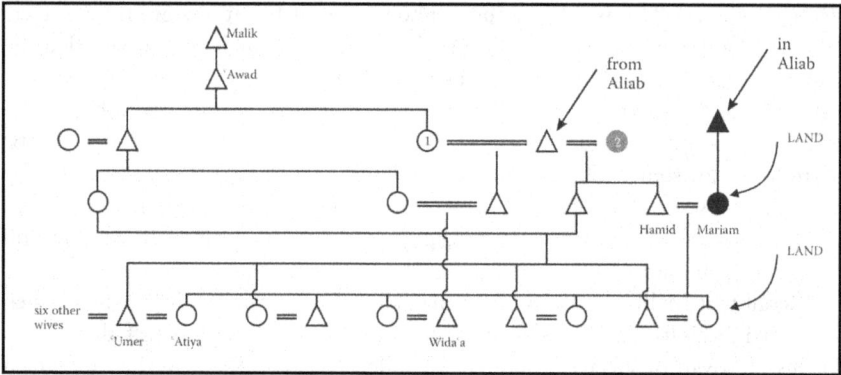

Figure 4.4 Incoming land generates lineage consolidation

Uterine Kinship and Endogamy

Women's second marriages also extend links that are consolidated in subsequent generations. A shared common grandmother (*haboobatum wahiida*) can also be the basis for a marriage between members of different descent lines. Such a couple would be described as *awlaad ʿamm*, on the understanding that their fathers are uterine brothers. Likewise, when a woman nurses another's children, the sharing of breast milk (maternal fluid) creates a sibling bond (*awlaad laban*) between her own offspring and the child she nursed and all its uterine siblings. While this has the immediate effect of prohibiting marriages among those children, it potentiates marriages among their progeny. Such marriages create the conditions for close kinship and future marriageability which, when acted upon, enfold and concentrate valued ties.

The pivotal positions of women in tracing kinship, devolving property and claiming moral affiliation are expressed in the 'lateral genealogies' kept by community sages. A segmentary lineal genealogy may list sisters and their various husbands for one or at most two generations, revealing how women provide the practical connections of participatory personhood, the flesh and blood that bind these variously disparate lines of descent. So tangled and knotted are 'descent lines' in northern Sudan that it's hardly surprising that British officials had trouble delimiting 'tribes'.

Conclusion

Shared bodily substance acquired from both male and female forebears provides a reference point for common identity and hence propriety and possibility for negotiating future marriages. Yet it does not preclude the limited extension of ties to non-kin. Hofriyati recognize a need to diversify their options as well as condense existing relations. Still, a disparate mix of ontological matter in one generation of siblings creates opportunities for marriages among their children that are desirably 'close'. In the ebb and flow of kinship, imbricated relations can be mobilized, built on or suppressed as circumstances and individual interpretations of these require.

Kinship in northern Sudan is about potential futures as well as possible pasts. Patrilineality provides a line of bearing by which an ever encroaching past transforms the discretionary cognatic present into an ideal 'Arab' pedigree. The messiness of the present is 'recast' as patrilineality by suppressing ideologically ephemeral female links in ascending generations. Practically significant groups are clusters of kin who may not share tribal affiliation or even a common name. Alliances between regions and clusters are maintained by successive close marrriages, and kinship calculations are both prospective and retrospective at one and the same time.

A last – hypothetical – observation: in contrast to alliance theory wisdom, it may be better to think of kinship as inclusive rather than preclusive among those who conceive of themselves as endogamous. This too may be counter-intuitive. Endogamy is conventionally thought to be restrictive, as closing off avenues of support while massing it in established relations. I suggest, however, that, because of their cognatic ambiguity, endogamous systems may be more responsive to changing circumstance than exogamous ones. Of course, endogamy and exogamy are relative terms; yet, to the extent that they reflect local ideologies, they are in contrast. Put simply, endogamy rests on a theory of sameness and exogamy on a theory of difference. Philosophical incongruity may be the operative distinction between the two.

Prescriptive exogamy requires that identities be restricted: perpetual relations between groups rest on their prior difference and the formal assignment of offspring to one group alone. Proscriptive exogamy is more manipulable, perhaps, but still depends on exclusionary principles. For, at some point in the calculation, those on one side of the ledger of relatedness become unquestionably unlike those on the other and thus potential mates. Here lineage affiliation notionally precedes marital alliance.

The logic of endogamy, however, seems less exclusionary than incorporative. 'Lineages' as such are the continuously accreting residue of memory, emerging after the fact, when marriage ties have been forgotten or suppressed. Descent and marriage are intimately linked and yet the order of their salience appears the inverse of that where exogamy is the rule. Instead of 'descent, therefore marriage', the logic is 'marriage, therefore descent'. Riverain Sudanese support selectively absorbing disparate others, condensing exchange, intensifying sociality and repelling moral and temporal dissolution by using present and past relatedness to create even closer ties one or more generations hence. Their assertions of sociality are performative and future-oriented and yet based on 'essential' attributes all the same.

Chapter 5

Descent and Descent Ideologies: The Blue Nile Area (Sudan) and Northern Kenya Compared

Günther Schlee

Introduction

Since Maine,[1] social and political theory has been pervaded by the idea of contract. Social relations are thought to be shaped by the intentions of people, by agreements between people, and by the interests they attach to these relationships. Constructivism has further directed our attention to the ideas people have about society and to the question of how social action is guided by such mental constructs. The merits of both contract theory and constructivism are beyond doubt. But have these dominant theories, with their focus on what happens in our minds and how we harmonize our ideas with each other, not led our attention away from a fact that should be too obvious even to deserve mention? Namely, that society is not just something that happens between our minds but something that also involves our bodies. Society is not just a way to think about people or a contractual form of interaction between existing people, but part of the way people are physically produced. It sets the framework in which we meet and copulate, reproduce and raise our young.[2]

Even in modern post-industrial countries, with their all-time low fertility, biological reproduction is still the most important form of recruitment into society. In agrarian or agro-pastoral settings one can even speak of reproduction strategies. Here, kinship and marriage, as the essence of politics, are a central concern of social actors.

At this point the argument could take a demographic turn. We could discuss different growth strategies of people and their herds and other assets, expansionist or more conservative, and I have indeed taken up this line of argument elsewhere (Schlee 1988, 1994a; see also Gray et al. 2003). Here, however, I want to explore different forms of identification and solidarity: the question of 'to whom children

---

1. Maine (1986 [1864]: 295ff.); see also Feaver (1969), Fox (1993: 96ff.).
2. This is true for historically older and younger social formations alike. There is much plausibility in Maine's thesis that 'status' had more importance in ancient times while 'contract' prevails in contemporary society, but this does not affect this argument about society setting the framework for biological reproduction.

belong' relates to principles of descent and descent rules; the question of 'how marriages, apart from producing a dyad of a man and a woman and a framework for biological reproduction, contribute to social cohesion' relates to the nature of affinal ties. I explore both these dimensions and processes, starting with descent.

Some people have rather purist ideas of unilinear descent. They claim to belong to the same group because they descend from a shared male ancestor exclusively through male links or from an ancestress exclusively through women. In the cases discussed here, we exclusively deal with the first of these two options, patriliny. The extent to which people actually organize themselves according to these ideas is a contested issue.[3]

Kinship in general, defined in terms of uterine[4] and affinal[5] links, plays a role in the definition of groups or the construction of networks in a variety of ways and degrees. Whichever form classifications using the kinship idiom may take, they may or may not be used in a metaphorical sense to postulate the unity of a group that is actually formed by co-residence and belonging to the same place. Also, groups originally constituted as local units may evolve through intermarriage into a dense web of kinship and may even adopt the ideology of being a unilinear descent group.

These interrelations between kinship and marriage, descent and locality are one of the classical problems of anthropology and, at the same time, they are one of the most up-to-date ones. This field comprises many unanswered questions, and new questions come up as the object under study undergoes change. The contexts in which people operate with notions of kinship and descent have become regionalized: they cross the rural/urban divide, and they have even become globalized. This causes adjustments to be made.[6] Also, and unfortunately for the people concerned, the effects of violence and disruption on social organization are a topic of increasing relevance.

The recent anthropological fashion is to drop the notion of kinship and to use a wider concept of 'relatedness' instead. This approach is not helpful at all. How is it possible to analyse how ideas and practices about kinship, marriage, belonging to a named group, belonging to a locality or bonds resulting from fostering or friendship interpenetrate each other conceptually and interact with each other in social reality if they are all referred to by the same fuzzy label from the start? Some analytical separation between concepts is necessary.

---

3. Often they do so to quite an extent. The deconstructionist critique, with excessive scepticism of models, overshoots the target by denouncing descent groups wherever they have been claimed to exist as anthropological fictions.
4. 'Uterine' means through a woman. Children of a male immigrant, whose own patrilineage is not locally present, may be counted as part of the patrilineal group of their mother. The resulting patrilineal genealogy would comprise one uterine link: So-and-so, son of So-and-so, daughter of So-and-so, son of So-and-so … If relationships through both parents are equally relevant on a regular basis in each generation we speak of cognatic kinship.
5. 'Affinal' means through marriage. In anthropology it replaces the colloquial English term 'in-laws', which has connotations that are not universally applicable.
6. Falge, in this collection, describes how the Nuer in Wisconsin, in comparison to the Nuer in Gambela, include a wider range of people in certain categories because otherwise they would be too few.

Unilinear Descent and the Segmentary Lineage System: Some Clarifications

The study of unilinear descent systems has suffered from a tendency to over-generalize from its beginnings in the nineteenth century. In modern understanding, (uni-)linearity is a way of passing on group membership and various types of goods and entitlements from one generation to another. Evolutionary myths about patriarchy and matriarchy mixed this issue up with questions of relative power or influence between the genders. This gave rise to popular assumptions that equate patriliny with male dominance and matriliny with female dominance, while the ethnographic data suggest a much more complex and differentiated picture. Some or most men may or may not have a dominating role in societies where goods and status are passed on along the patriline; the same is true of the rather loose association of the status of women and men to matrilineal modes of succession and inheritance.

In the case of matriarchy, it now seems to have been understood that matrilinearity might give women a structural advantage only in combination with matrilocal residence.[7] Otherwise there is no close relationship between matrilinearity and the power balance between the genders.

The term patriarchy has not been discredited to nearly the same degree, however. More people believe patrilinearity to be synonymous with patriarchy and to be equivalent to male dominance, and yet the variation in relations between men and women among different so-called 'patriarchal societies' is high: at the very least the implications of patrilinearity and the concept of patriarchy need more critical investigation.

I am going to speak about systems of alliance in the two senses of the term: affinal relationships as in 'alliance theory' and military or political alliances. The image of gender relationships conveyed to us by alliance theory in its classical shape, for example by Lévi-Strauss and his followers, is that of men sitting in the shade discussing the transfer of rights in women from one of them to the other. Males in alliance theory consist of 'wife givers' and 'wife takers'. Before leaving the topic of patriarchy altogether, I want to make clear that this is an inversion of how things really work in many cases. Among Sudanese Arabs (Boddy 1989) the typical matchmakers are grandmothers and, to a lesser degree, mothers.[8] It is females who determine social and economic fates and chances of reproduction by deciding who marries whom. This is despite the fact that the society's official self-description consists of long genealogies that list names of men only: the sons of Ahmad

---

7. In avunculocal settings, where property is transferred between men (mother's brother (MB) to Ego) and wives are outsiders joining related groups of men, there is no reason to expect their position to differ much from patrilineal and virilocal settings.
8. For female agency in connection with marriage in Southern African cases, see Kuper (1982). Here a woman asks her brother for one of his daughters for her son to marry, arguing that her brother was enabled to marry by the cattle paid as bridewealth for herself, thus owing his daughter's existence to her. The title of the book, *Wives for Cattle*, sounds like women being transferred against payment. But the content makes quite clear that such a view does not always correspond to who actually transfers whom.

Suleymaan Abu Bakr splitting from the sons of Mahmuud Suleymaan Abu Bakr and so on. This is not to brush the issue of discrimination against women aside. Younger women might be restricted in many ways, and they might have as little free choice in marrying as they do in many other issues, including education and work. Wealth and prestige might be in the name of men. But the fact that it is often through the agency of senior women that decisions are made about who marries whom, and thus about to whose descendants wealth and hereditary status go in the next generation, seems to me, at least, to necessitate an important modification to popular theories about 'patriarchy' and to alliance theory.

This revised 'patriarchy', in which many of the patriarchs are in fact matriarchs and many of the patrons matrons, might still be a far cry from Western ideals of free choice for a young women and the love marriage, but among the people described by Boddy it is other women and not grey-bearded men, as suggested by the older alliance theory and gerontocratic models (Spencer 1965; Meillassoux 1991), who make decisions about young women and young men. The love marriage appears to be a fairly recent invention even in the West and, if you look at divorce statistics, not a particularly successful one.

I shall discuss structures of descent by using diagrams. Some of these structures involve groups defined by relationships between males only. This is not only inevitable because these are the relationships that are remembered, but even justifiable because these are real structures and relevant to action. It is irrelevant to these structures whether they go back to the strategic actions of men or women. In this they are like configurations on a chessboard: after given moves you get a certain configuration, irrespective of whether the players are male or female. But, when shifting back to a perspective of action, we should be prepared for the actors to include more women than expected.

I have started by saying that the study of unilinear descent systems has suffered from a tendency to overgeneralize. This also applies to the most famous model derived from this line of study, the segmentary lineage system. From the paradigmatic cases, the Nuer (Evans-Pritchard 1940a) and Tallensi (Fortes 1969), this model spread to be applied to one of what was believed to be two basic types of society in Africa (Fortes and Evans-Pritchard 1940) and beyond, like the Mae Enga of New Guinea (Meggit 1965). Then a reaction set in. The model was said to overstress certain aspects of the paradigmatic cases and to ignore others. There is an abundance of secondary writings about *The Nuer*. That anywhere there was ever anything that might with some justification be called a segmentary lineage was thrown into doubt. A strong interpretation of what Adam Kuper says on the subject in *The Invention of Primitive Society* (1988) might suggest that such lineages have only existed in the minds of anthropologists. He himself, however, told me that he adheres to a weaker version of this position: he only wanted to warn against overgeneralizations. The segmentary lineage model fits some societies better than others.

In addition, the segmentary lineage system, being very, very structural in the models used to describe it, is simply no longer fashionable in this post-structuralist age. Take the example of Somali studies. In Somali society lineages are all-pervasive. The segmentary lineage model, with some modifications, already applied by I.M.

Lewis (1999 [1961]), might fit the Somali even better than the Nuer, although the Nuer are the paradigmatic case for this model. It is therefore most peculiar to watch students of Somali society bend over backwards to find something different from segmentary lineages simply in order not to be accused of applying this old-fashioned model (Schlee 2002).

Subtypes of Patrilinear Descent

This chapter is not a case study that focuses on one ethnic group. It compares two regions, one in the Blue Nile area of Sudan, the other in northern Kenya, both of which are pluri-ethnic settings. In these settings, processes of group integration and inter-group articulation, among them ethnogenesis and change of ethnic identities, have been studied.[9]

To be fruitful, a comparison must establish a framework – some broad dimension of sameness that extends to all cases under examination – and then go on to look at the differences within this framework. This particular study is about patrilineal descent reckoning and some of the differences between its many types.

The Cushitic-speaking pastoralists of northern Kenya and the inhabitants of the riverine oasis of the Blue Nile in Sudan share the basic feature of their descent reckoning: it is patrilineal, at least as far as the outwardly demonstrated, 'official' forms of belonging are concerned, as they are, for example, expressed in naming (patronyms or lineage names function similarly to surnames in Western societies). Within the framework defined by this general shared feature, however, they show many contrasts. Our main finding will be that similarity to the extent that descent reckoning is patrilinear does not preclude the possibility that, in different patrilinear settings, group structures, perceptions of sameness and difference (social maps, identification) and binding forces (integration (psychosocial and/or systemic), solidarity, ties (cross-cutting or not) differ widely.

While in the riverine Sudan we find patrilines that undergo lower-level segmentation in each successive generation, the Somaloid-speaking Rendille and many of their Oromo-speaking neighbours in northern Kenya have clans that do not have a high rate of change over time and are, in emic theory, even regarded as pseudo-species: units that are set apart from each other by biological characteristics.

If we look at marriage patterns, we find that first-cousin marriage is permitted, or even preferred, in Sudan, while among the Rendille, to take the extreme counter-

---

9. Research in northern Kenya has been conducted since 1974 and has involved extensive study of Cushitic languages (Rendille, Oromo and, to some extent, Somali) in addition to Swahili, to enable the researcher to record clan histories and interactions from the perspectives of the different groups involved. The time spent in northern Kenya over the decades has added up to more than eight years. Research in Sudan started relatively recently, in 1996. Due to other commitments over this time, stays in this new research setting have been shorter (just a month or two at a time) and the researcher's linguistic knowledge of the region has remained more limited. He has some knowledge of Arabic and little of the West African languages spoken among some of the immigrants to the Blue Nile valley. For certain aspects of this research he relies heavily on the help of colleagues, such as Al-Amin Abu-Manga (who also has a contribution in this volume) and Awad al Karim.

example, there is not only clan exogamy, which extends to adoptive 'brother' clans and equivalent clans in neighbouring ethnic groups, but also a prohibition against marrying from one's mother's clan and from one's father's mother's clan. Rendille clans can therefore never develop or preserve any discontinuities in physical type or other biological features because there is a constant gene flow through intermarriage between them. Paradoxically, as I will show, the Rendille themselves believe the opposite and ascribe not only different temperamental dispositions but also contrasting biological features to their clans. On the other hand, among the riverine Sudanese, where endogamy enables relatively small groups to be independent from each other in their biological reproduction for extended periods of time, we find, again somewhat paradoxically, adherence to a religion with a universal message and a relatively weak development of exclusivist group ideologies.

In her contribution (this volume) Boddy has put forward the point that, '[t]o put it simply, endogamy relies on a theory of sameness and exogamy on a theory of difference'. This point sounds counter-intuitive, since one would expect attitudes that one might call 'micro-racialism(s)' among closely intermarrying people who duplicate and triplicate the same alliances and – from a biological point of view – share more genes with each other than with the surrounding people. One would expect more inclusive views about who belongs to one's own people among those who belong to large exogamous groups and intermarry cyclically, or in whichever sequence, with other such large exogamous groups, who have a universe of real and potential affines of several thousand people. Well-established anthropological theories would support this expectation.[10] But maybe our data do not, and would rather support Boddy's view: exogamy can go along (or even regularly does go along) with an ideology of difference. Having a wide universe of potential marriage partners

---

10.   In a summary of anthropological theorizing about marital alliances, Holy (1998: 124ff.) draws a line from Tyler to Lévi-Strauss. Alliance theory stresses the 'significance of marriage alliances for the cohesion of society'. The stress is always on the widening of networks, not on making them denser. This can be illustrated with Lévi-Strauss's explanation why, among the two types of unilateral cross-cousin marriage, the one with the MBD (mother-brother's daughter) is more widespread than the one with the FZD (father-sister's daughter). Marriage with the patrilateral cross-cousin 'precipitately closes the cycle of reciprocity and consequently prevents the latter from ever being extended to the whole group' (as marriage with the MBD would allow). In other words, the wider alliances that encompass more descent groups are thought of as being able to integrate whole 'groups' or 'societies'.

Certainly there are merits in alliance theory, but some caveats are also in place. Intermarriage does not invariably lead to cohesion or integration. We should not forget the many instances in which 'those whom we marry are those whom we fight' (Lang 1977). A strong value attached to exogamy leads to potential marriage partners and potential enemies coming from the same category of people, namely 'strangers'. The idyllic view the Arapesh express of brothers-in-law, when they say 'with whom will you hunt, with whom will you garden, whom will you go to visit?' (if you do not have brothers-in-law), might not be shared by people whose relationships to their affines oscillate between war and uneasy peace (the Arapesh are cited by Margaret Mead, who in turn is cited by Lévi-Strauss 1969: 485 and Holy 1998: 126; on exogamy and cross-cutting ties in general and war, see also Schlee 1997: 577f.).

does not mean that feelings of sameness and solidarity are extended to all these people. This can be further illustrated by the tendencies of fission among the Rendille of northern Kenya. But first we turn to Sudan.

The Blue Nile Area

Not unlike the lower stretches of the White Nile and the valley of the combined Nile below Khartoum, the Blue Nile area of Sudan is a favourable habitat that has always attracted immigrants. Irrigation is practised along the river and, since dam construction in the 1920s, on vast irrigation schemes fed by channels that reach out far into the surrounding plains. From Sennar (Sinaar) southwards rain-fed agriculture also plays a major role, increasingly so as one moves south.

The research comprised some wider travelling along the river, mostly with Al-Amin Abu-Manga, collecting oral traditions, and a village study of Barankawa supplemented by visits to other locations within walking distance or within the reach of tractors and pickups from the University of Sennar, Abu Na'ama, where I was stationed most of the time. In this area, on the left bank south of Singa, almost everyone seemed an immigrant of some sort. Very few people identified themselves as 'Funj', thereby claiming an affiliation to the Funj Sultanate, the state dominant in the area before 1823. The wider research question, not unusual in ethnicity studies, was phrased in Barthian terms: if ethnicities articulate at the boundaries with each other, in this setting, where everyone is a migrant and all boundaries are new, new forms of ethnicity can be expected to develop. What forms do these take?

Most of my interlocutors were Arabs or 'Fallata'. The Arabs of Barankawa were made up of recently sedentarized Kenaana cattle-owning agro-pastoralists (in 1996 I still found a cluster of them living in tents in the vicinity of relatives who had built mud-and-thatch houses earlier) and Jaᶜaliyiin. The latter here had intermarried and lived interspersed with Rufaᶜa Arabs.

'Fallata' is an extremely heterogeneous category. It comprises the descendants of high-status West African Qur'anic scholars, often Fulɓe, who came to the Nile centuries ago. It further comprises aristocrats who came from West Africa as a part of the hegira, the withdrawal of the believers from the expanding colonial rule in Nigeria and French West Africa in the early twentieth century: these included, among others, descendants of Shehu Osman dan Fodio and Hajj ᶜUmar Tall, who live along the Blue Nile. The 'Fallata' also comprise the descendants of poor pilgrims who did not make it back to West Africa or who decided that life was better here. The overland route from West Africa to Mecca was made obsolete only by the decreasing cost of air travel in the 1970s (Birks 1978). We also find ordinary labour migrants. The categories 'pilgrims' and 'labour migrants' overlap, since pilgrims were forced to beg or work to carry on with their journey, which might take years with these interruptions, and people primarily motivated by the search for labour might combine their journey with a trip to Mecca in the haj season. Apart from Fulfulde, Hausa and Kanuri are prominent among the West African languages spoken along the Blue Nile.

West African settlers are found all over northern Sudan (in the political sense) and beyond it. On a recent visit to Ethiopa I found a hamlet of Hausa speakers on the Dabus, a tributary of the Blue Nile in the Beni Shangul area. They had come from Gedaref in eastern Sudan. Nomadic movements of Fulfulde speakers are discussed in another contribution to this volume (Dereje Feyissa and Schlee).

In Barankawa village, the West African element is represented by a community of Kanuri speakers, who are referred to as 'Bornu'. Many of them, mostly the young girls, work as household helps for the academic staff of the Abu Na'ama campus of Sennar University. This campus, locally just called *kulliya*, 'the faculty', comprises the head administration of the university, which has other faculties in Sennar and Suuki, and the Faculty of Agriculture with its many demonstration and experimental fields and its cattle and poultry. In the misery of the war economy of the late 1990s, these activities helped the academics to survive. This institution requires much manual labour, which is recruited in the surrounding villages and comprises Bornu and Kenaana.

Kenaana, Ja^caliyiin and Bornu are perceived as the main groups who make up the population of Barankawa village. They live in tribally[11] largely homogeneous neighbourhoods and are thus clearly visible as separate communities. In each neighbourhood a full micro-census was made of a spatially contiguous line or cluster of houses and genealogical data were collected as comprehensively as possible. The genealogical data reflect the history of immigration.

The Kenaana, former mobile cattle pastoralists, have not settled here for long, but their cohesion pre-dates their sedentarization. They are related to one another and have been in contact with one another in other locations before. Going upwards (i.e. back in time), from the generation of the living adults (i.e. the informants), one finds the closed loops typical for the prevalence of marriage with kin in the ascending generations. This is because different kin types converge on the same person: e.g. father's father's father and mother's mother's father are the same person, if father has married his father's sister's daughter, the mother of Ego, etc. In the descending generations the same loops occur, if the descendants of Ego intermarry (see Figure 5.1).

To illustrate the 'loops' mentioned in the text, one such loop, containing a smaller one, has been highlighted in Figure 5.1 in the ascending generations, and one of several marriages between the children of the brothers ^cAbdurrahman and Muhammad in the generation descending from that of the informants.

The Ja^caliyiin of Barankawa are descended from pioneer traders who came from Dinder, well to the north of the confluence of the Niles. These traders in slaves, ivory and cotton settled here in the early nineteenth century. They were young men who married women from the nomadic Rufa^ca Arabs. At a later stage they abandoned their families and went back to Dinder to conclude a proper marriage there, i.e. one with a cousin. The origin from mixed marriages is reflected by both the residential patterns and the marriage patterns of the present-day Ja^caliyiin of Barankawa: they are the only group among the three groups under examination that do not live in homogeneous settlement areas, but are interspersed with Rufa^ca. In some of their marriages the old alliance with Rufa^ca is renewed; other marriages are between Ja^caliyiin (Figure 5.2).

---

11. 'Tribe' and its Arabic equivalent, *qabiila*, pl. *qabaa'il*, do not have the pejorative connotations in Sudan and the whole Middle East that they have in most of sub-Saharan Africa.

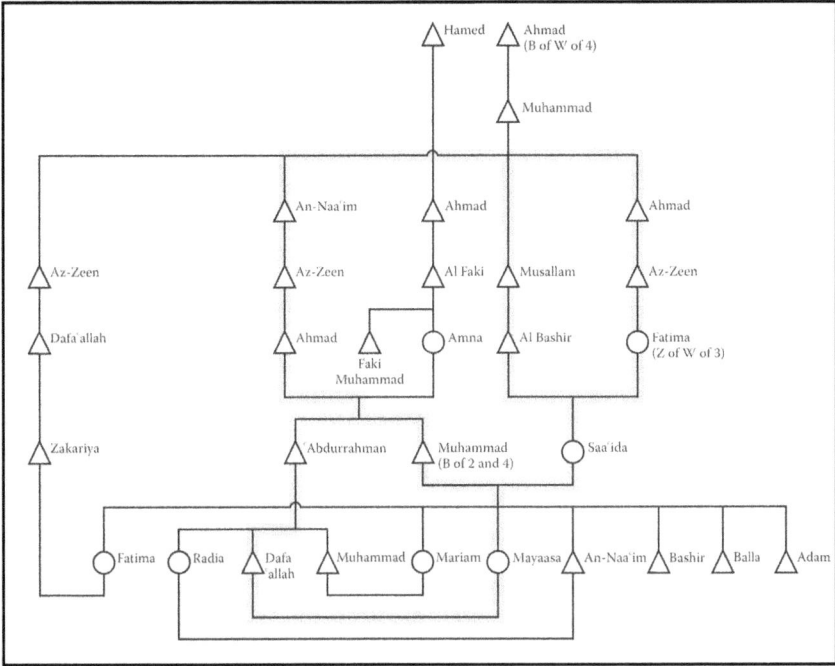

Figure 5.1 Kenaana, Barankawa, neighbourhood 5
(The numbers are cross-references to genealogies of other Kenaana households, which are not depicted here.)

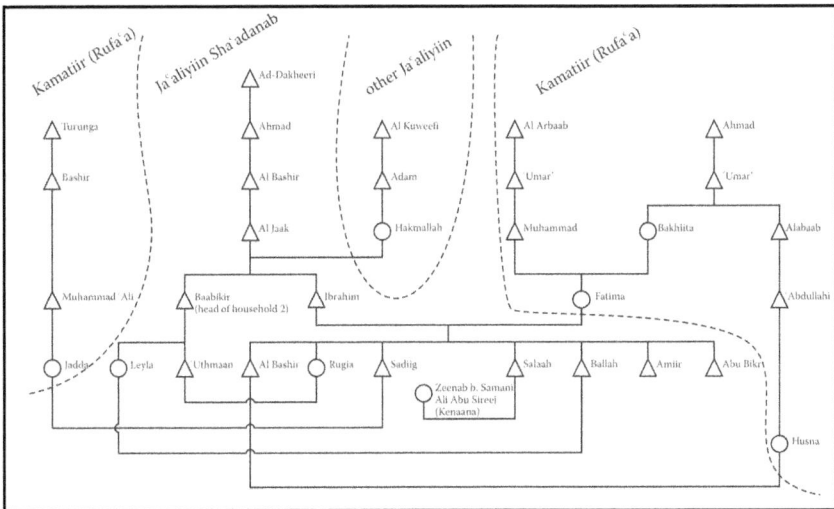

Figure 5.2 Jaᶜaliyiin, Barankawa, household 6

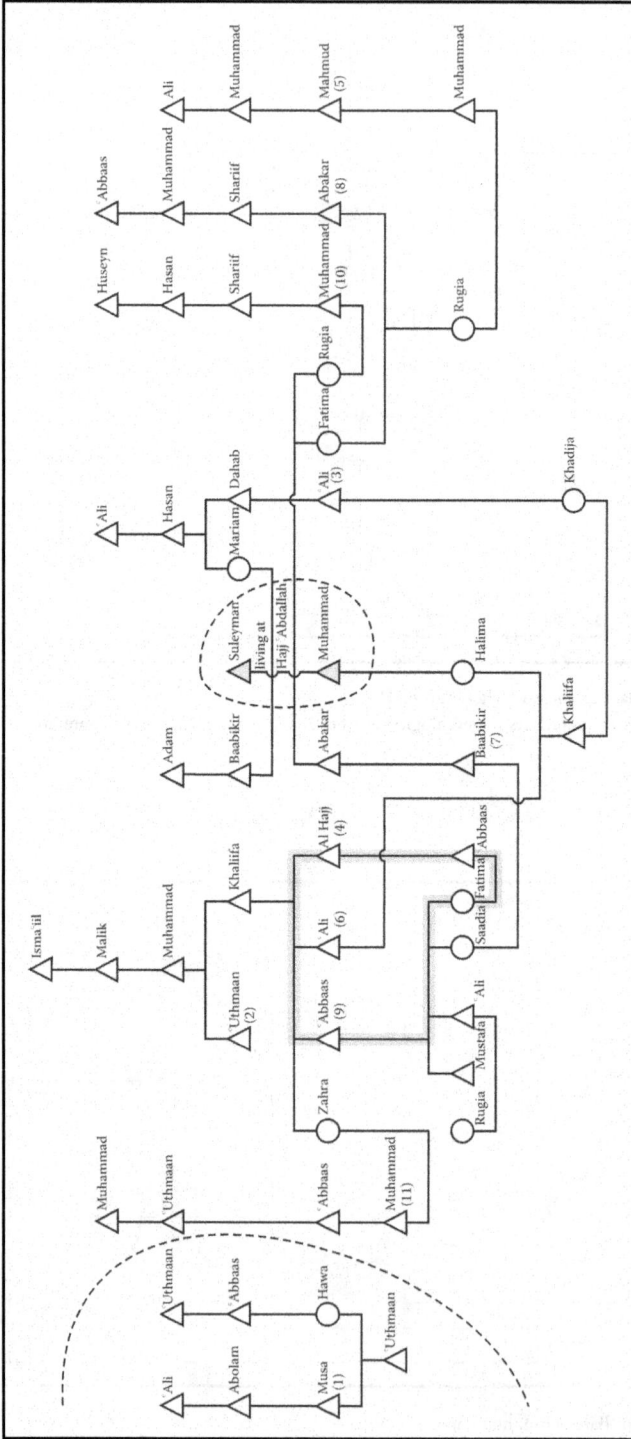

Figure 5.3 Bornu, Barankawa, relations between households
(The numbers are assigned to the heads of households documented in the census with corresponding numbers.)

The genealogies of the 'Bornu', descendants of Kanuri-speaking pilgrims from what is now Nigeria, reflect clearly the fact that these migrants have arrived here as individuals or as fragments of nuclear families (a mother with her son or the like), not by the coordinated movements of larger kin groups. Going upwards in the diagram, the lines of descent separate. In the younger generations, however, we find the loops indicative of kin marriage also among the Bornu. This means that these Kanuri have started to behave here in Barankawa like one endogamous 'tribe'. This, of course, does not preclude the possibility that in their earlier places of residence they practised similar patterns of marrying 'in', but, if they did so, it was with other Kanuri, not with the ancestors of their present neighbours. In the Kanuri case it is quite clear that their endogamy is partly involuntary. If the other groups did not ascribe a low status to them, there might be some intermarriage (Figure 5.3).

Northern Kenya

In what I am going to say about patrilineal societies of North-East Africa, the segmentary lineage system finds its place among other descriptive models. I try to avoid the past mistakes of privileging it beyond measure or of discarding it altogether.

While the Somali have a segmentary lineage system that seems to fit the classical model even better than the paradigmatic cases, the Rendille camel nomads of northern Kenya, who speak a closely related language, deviate from this model in a number of ways. They have nine clans, not counting the newcomers, Odoola, who might have arrived in the seventeenth century. The Rendille still say that they have nine clans, and with Odoola they are ten. There are also other cases in which it can be shown that two or three centuries are not enough to become fully Rendille. Processes of fusion and integration are slow, and so are the processes of fission that are so characteristic of the Somali segmentary systems, in which small units grow into large ones and their component parts take over the functions of the smaller units. The Rendille attribute personalities and even biological characteristics to their clans (see the section on 'Ethnobiological categories' in Schlee 1994c). There is a clan with a special affinity to water, who are said to have originally crossed Lake Turkana to join the Rendille. Their daughters need to be forgiven if they do not wait for their turn at the well; they just cannot resist the temptations of water. The daughters of some clans are believed to have a pregnancy length of ten months while those of other clans give birth after nine months. It is the daughters that pass on such clan personalities. Rendille ideally live in patrilocal clan settlements so that co-resident males mostly share patrilineal descent. The differences of temper and character between them are therefore explained by the different origins of their mothers.

Such beliefs are, of course, contra-factual. The marriage rules that necessitate far-reaching exogamy can only result in a homogeneous mix of genes. Any clustering of biological traits in certain clans is quite impossible. But, contra-factual or not, such beliefs can only be maintained if clans have a certain time stability. If the clan map changed after two or three generations, they could hardly be believed to be what we would call natural and given. And that is what the Rendille claim: the clans exist from the time God allowed people to come out of the ground, and they are believed to be universal: Rendille have asked me whether the Germans have the same nine or ten clans.

The cousin terminology of the Somali (Figure 5.4) is the same as that used by anthropologists who want to distinguish all kin types, i.e. who instead of lumping together different types of relatives under labels like 'uncles' or 'cousins' distinguish them all and speak of father's brothers (FB), mother's brothers (MB), father's sister's husband (FZH), mother's brother's son (MBS), etc. This type of terminology is called 'Descriptive' if the etymologies are clear, i.e. if the MB is called 'mother's brother'. It is called 'Sudanese' if the terms have the same specific meanings but cannot be analysed. In Arabic, for example, there is a term *khaal*, which means MB but does not contain the words 'mother' and 'brother'. This term is a different root and not a combination of semantic elements that can be separated by analysis. The Somali kin terminology contains Descriptive and Sudanese elements. The term for MB, *abti*, is non-analysable while the term for MBS, *ilm abti*, can be translated word by word and is then found to mean just that: 'son of the *abti*', 'son of the MB'. Descriptive and Sudanese-type terminologies are not infrequent in Sudan, as the name suggests, and in the Middle East in general. Arabic, for example, has a kin terminology of this type. They are frequently associated with segmentary lineage systems.

The cousin terminology of the Rendille, in contrast, is of the Omaha type, which reflects a patrilineal clan ideology very well (see Figures 5.5 and 5.6). Some terms the Rendille use have an etymology that makes the closeness of this relationship between the kinship terminology and the social organization of the Rendille particularly clear. A sister's or clan sister's child is *eysim*, the same as a remainder of milk left in a

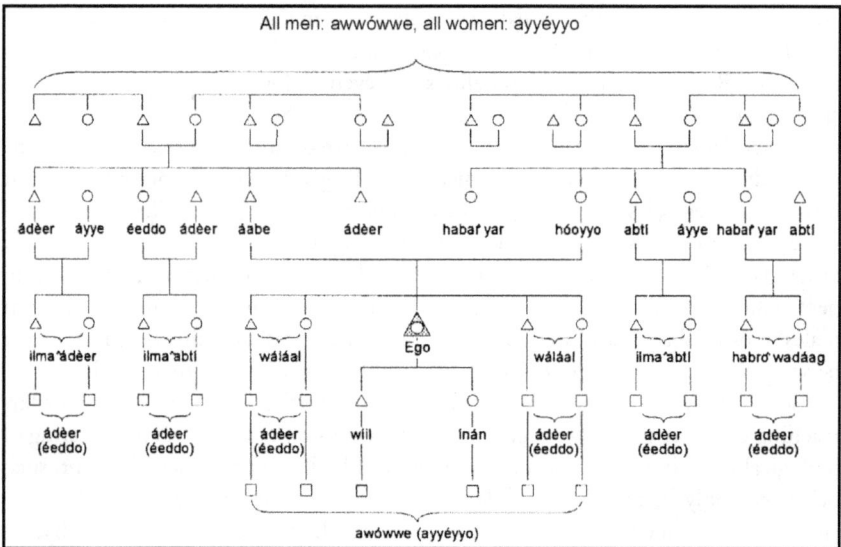

Figure 5.4 Somali – male Ego (female Ego) (source: Schlee 1994b: 383, Plate I)

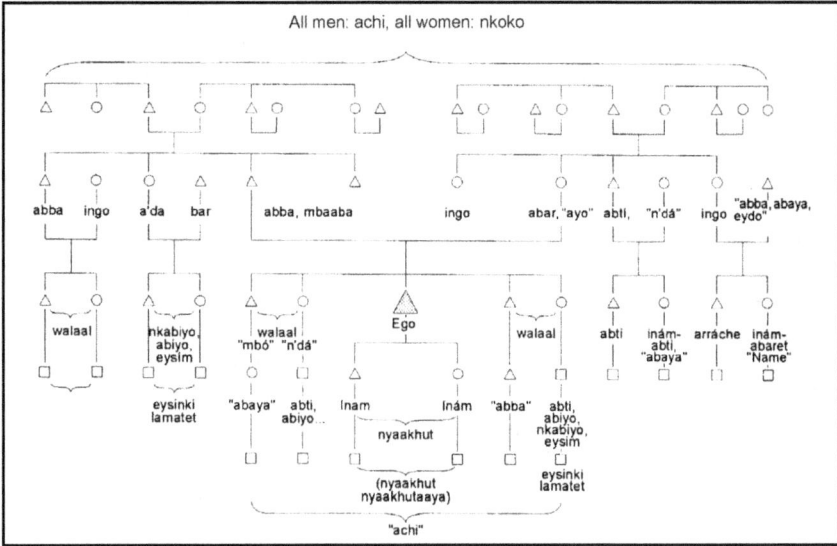

Figure 5.5 Rendille – male Ego (source: Schlee 1994b: 384, Plate II)

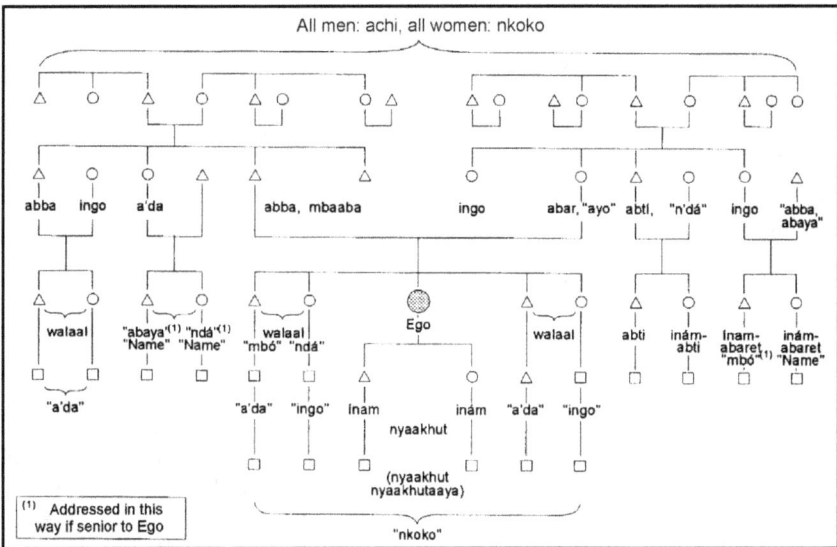

Figure 5.6 Rendille – female Ego (source: Schlee 1994b: 385, Plate III)

container. They are what is left of clanship once the clan boundary is crossed. We have seen that, according to the pseudo-biological convictions of the Rendille, children of the daughters of a clan exhibit characteristics of the clan of their mothers in a marked way, but, as the descent reckoning is patrilineal, they do not pass them on to their children. These would be *eysim ki lamatet* or second-degree *eysim* of the clan of their FM, which would preclude intermarriage but has little importance beyond that, while the relationship between an individual and his or her mother's patriclan and the customs associated with it would deserve a separate chapter.

A look at the cross-cousins reveals the defining feature of an Omaha-type terminology. Cross-cousins are those linked to Ego by persons of different gender in the parental generation, i.e. the children of the FZ and the MB. In an Omaha terminology MBC are equated with someone in an ascending generation, while the FZC are equated with someone in a descending generation. In fact, we find that the MBS is called *abti*, like his father, Ego's MB. The FZC, on the other hand, the patrilateral cross-cousins, are referred to by the same term as the ZC. In a way, they are 'nephews' and 'nieces', being classified in the same categories as these relatives of a junior generation. They are the *eysim*, already mentioned: the children of a daughter of Ego's clan.[12]

To explain the marriage rules associated with this kinship system we have to leave the Ego-centred-kindred perspective expressed by this diagram again and regard the matter in terms of groups. Ego is not allowed to marry from his own clan or from the clan of his mother or that of his father's mother. Marriage with a girl who originates from Ego's father's father's mother, a classificatory FFMBSSD, is, however, not only allowed but preferred, especially in the case of firstborn sons, who are those who ritually matter. The preferential marriage rule can thus be expressed by this formula (left of Figure 5.7):

The diagram in Figure 5.8 depicts the overlapping loops that result if this rule is put into practice. It depicts the marriages of members of one patriline, A. If we include the marriages of all male members of B, C and D, we would get many more such overlapping loops. I have refrained from drawing all these loops and leave them to the reader's visual imagination. It is best to think of a plate full of spaghetti: a dense mesh of overlapping loops, not very easy to discern. One of the best attempts at capturing this complexity on paper can be found in Adam Kuper's representation of the nearly identical Tswana system (Kuper 1982).

---

12. It is striking to see how different the Somali and Rendille are in kin terminologies and social organization despite the fact that they speak the same language. In another paper (Schlee 1994b) I argued that the Rendille might once have had a Descriptive terminology as well, but have been transformed into an Omaha one under Nilotic influence. There is plenty of evidence of Maa influence on Rendille culture. At present, the Rendille and the Maa-speaking Samburu are 'nomads in alliance', as Spencer called his 1973 book. Before, they might have had similar relationships with other Maa groups. The diagrams included in Schlee (1994b) summarize different accounts of Maa terminologies (own material on Samburu; Hollis 1910 and Merker 1910 on Maasai). The reader who has followed me to this point will by now be able to trace the Omaha elements in them.

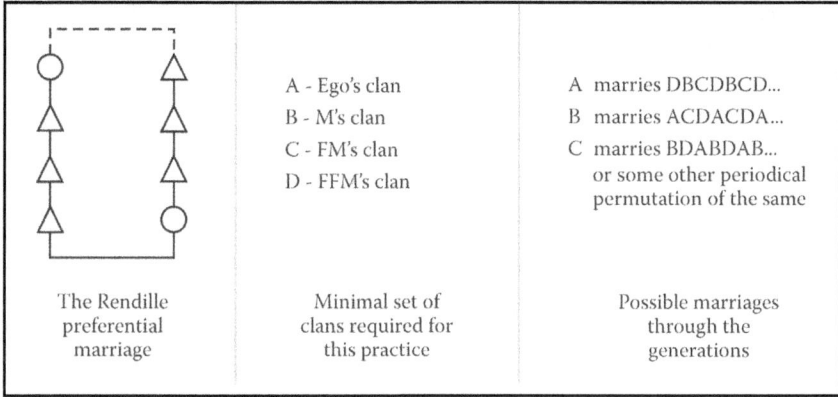

Figure 5.7 The Rendille marriage
(Explanation: M = mother, FM = father's mother, FFM = father's father's mother)

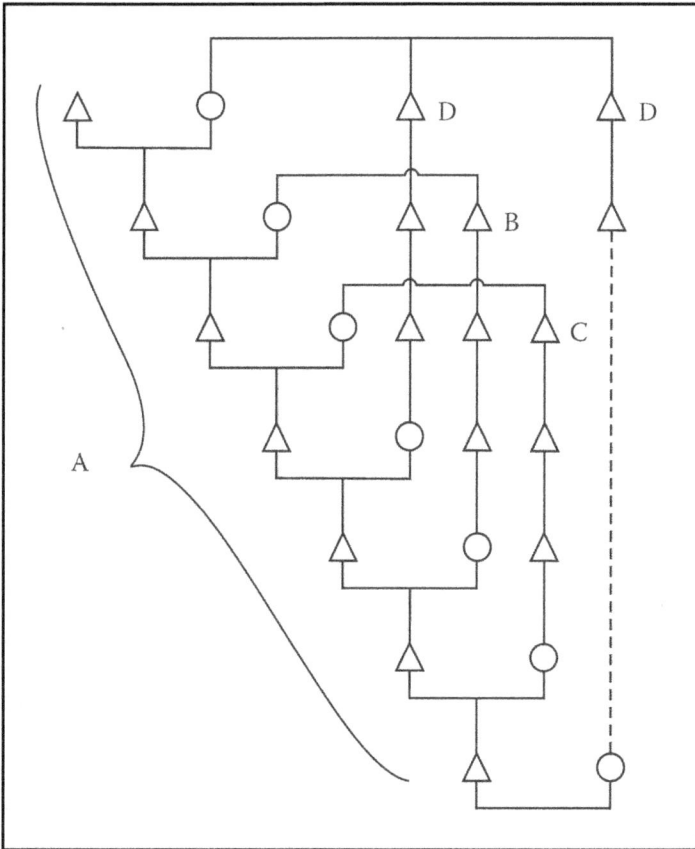

Figure 5.8 Cyclical marriage of the patriline A with the patrilines B, C and D

A recurring marriage with the same clan, of course, does not require a brother-sister link, since the rule is not about marrying an FFMBSSD in the strict sense but a girl from the FFM's lineage. An FFMFFBSSSSD, an FFMFFFBSSSSSD and many others would do. Also the number of generations up to the sibling link does not need to be the same as that down again. A FFMFBSSSD, for example, although by English reckoning a distant 'aunt', would do, if of approximately the right age. The rule is simply to marry from the FFM's people.

However, we should not forget that this is the minimal fulfilment of Rendille exogamy rules. A given patriline can also intermarry with more than three other clans and the different patrilines of a clan with their different histories of intermarriage have such relationships with all other Rendille clans except those to whom adoptive links of brotherhood exist, a matter that would lead us far beyond the present topic.

Another matter that I cannot fully describe here relates to the inter-ethnic clan relationships, about which I have written a book (Schlee 1994a [1989]). Suffice it to say here that, for historical reasons beyond the scope of this chapter, processes of ethnogenesis in this area have repeatedly cut across clan boundaries, so that today we find the same clans among peoples of quite different linguistic and political affiliations. The Rendille share clans with the Gabra, Sakuye, Garre and others (see Figure 5.9).

Clan brothers are clan brothers whether you have ever seen them or not. They may speak a language you do not understand, and they may even be at war with you. The sisters of your brothers are your sisters, so you cannot marry them. If your mother's or your father's mother's clan has an equivalent in another ethnic group, the male members of these clans are your maternal uncles and the females forbidden to you as marriage partners. I once met a young Rendille teacher in a Gabra town. He was in love with a Gabra girl and wanted to marry her. By that time I had gathered some knowledge of clan history and was able to inform him that the girl belonged to a Gabra lineage that was considered to consist of brothers of another lineage, part of which had been taken as war spoil by the Rendille some 200 years ago and had evolved into a sub-clan of his own clan. Therefore the girl was his sister. The young man, who had been in a good mood up until then, became quiet and very serious. After gathering his thoughts, he said: 'Günther, please, please shut up. Never say a word about this.' It was clear to him that my derivation of sisterhood was entirely consistent with Rendille reasoning and that this relationship, if brought to the knowledge of the elders, was an impediment to marriage. I did shut up and it was for different reasons in the end that he did not marry the girl.

All of these marriage constraints exist equally for young girls and women and can be seen from their perspective: if a mother's or father's mother's clan has an equivalent in another ethnic group, all the male members of those clans are forbidden as marriage partners. The extent of people forbidden in marriage stretches far and wide. Evidently, and to summarize up to this point; clan exogamy and marriage rules of the type often associated with an Omaha kinship terminology draw many groups of people, each of which may be quite populous, into an individual's realm of action. In combination with inter-ethnic clan relationships, which appear to be more frequent than we used to think, they open a social universe. The

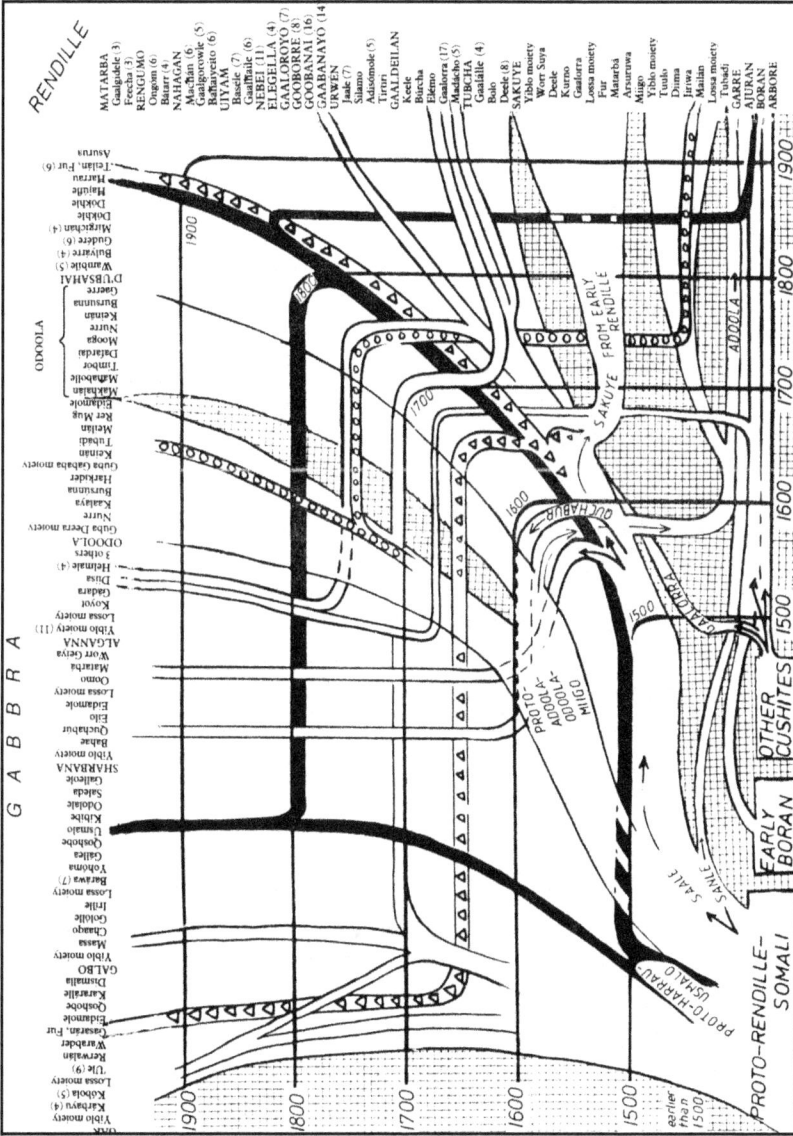

Figure 5.9 Links between clans in northern Kenya (source: Schlee 1994a [1989]: 235)

relationships generated by such a system may not be quite as universal as the Rendille believe when they assume the clans also exist in Germany, but they do imply relationships that criss-cross North-East and East Africa.

This stands in sharp contrast to the usual Middle Eastern type of patrilineal organization that is associated with Descriptive or Sudanese types of cousin terminology and in which marriage with the patrilateral parallel cousin and other cousins is allowed or even preferred. In societies to which this applies, a man and his brothers with their wives (who may be their cousins) are a perfectly self-sufficient reproductive unit. Their offspring can intermarry indefinitely without ever needing outside links.

Exogamy and Solidarity

Patrilineal descent with clan exogamy and patrilineal descent without clan exogamy are socially totally different things. Alliance theorists tend to assume that far-reaching systems of exogamy create wider solidarities and societal cohesion. The Rendille have exogamy rules that can indeed be called far-reaching because they preclude marriage not only within a clan, including its representations in other ethnic groups, but also with certain categories of uterine relatives and with the entire clans of these relatives, plus any adoptive or other brother clans these clans might have. The shortest summary of all this would be that the Rendille marry out-out-out. In spite of this rather extreme form of marrying out, the Rendille would be a bad example to illustrate the underlying assumptions of alliance theory. No matter how good the individual relationships between a man and the brothers of his wife may be,[13] and irrespective of the importance of the link to one's mother's brother(s),[14] in Rendille clan politics the multiple inter-clan marriages (i.e. the alliances in the sense of kinship theory) do not provide stable political alliances. The Rendille show clear tendencies of fission along the moiety line, in spite of the multiple uterine links across this line (see the conflict about the rituals leading to the marriage of the age-

---

13. These relationships are often quite good. Many men prefer to live with their affines rather than with their own clan brothers, with whom conflicts about property rights tend to lead to friction. Rendille clan settlements are clan settlements in name only and as far as the largest single component is concerned. Apart from a nucleus of people from the same clan, they may comprise a significant proportion of in-marrying males.

14. The importance of the relationship to one's MB cannot be overestimated, and not just because of the ritual property transfers (when Ego undergoes circumcision or has killed an enemy) it involves, but also because it provides a link to a senior male who is not burdened with paternal authority and filial submission. To underline the importance of the nephew (*eysim*), Rendille tell the following anecdote.

    A man was walking along a path in the company of his son and his nephew. The small group was intercepted by enemies. They asked the man: 'Whom shall we kill, your son or your nephew?' The reply was: 'Kill my son and spare my nephew. I can always beget another son, but there is no way to beget a nephew.' (Heard from Baarowa Adicharreh.)

set Ilkichi described by Schlee 1979). Boran[15] and Somali[16] examples could also be added to illustrate the fragility of uterine links when clan or moiety rivalry sets in.

Exogamy widens the social universe and it spreads social links. In this point alliance theory is right. But solidarity and 'cohesion' are not always part of these links, because the links are still perceived as links to 'outside'. Is, then, exogamy based on a theory of difference and endogamy on sameness, as Boddy so aptly put it? After all, repeated first-cousin marriage between the descendants of an immigrant and the descendants of the siblings of his wife will produce a degree of closeness that participation in a rotation of marital exchanges between large exogamous clans who marry 'strangers' will never achieve. Closeness here can be understood as biological closeness or belief in biological closeness;[17] it can result in or result from repeated cooperation and feelings of affinity or identity. How these components of closeness are interrelated needs to be examined on a case-by-case basis.

---

15. Electoral politics in Moyale (Kenya) regularly divides the Boran population, often bitterly, along the moiety line (Sabbo vs. Gona), although there is moiety exogamy and Sabbo men have mothers from Gona and Gona men mothers from Sabbo (Schlee and Shongolo, in preparation).
16. In spite of the importance of marital links between men and women, the possibilities that they hold for creating positive Somali clan politics and inter-clan conflict resolution through building on uterine links are questionable. I have heard rumours (and they might just be rumours but still they indicate mistrust) that in escalating conflicts people have killed their nephews who happened to live with them because they belonged to an enemy clan.
17. Even if we leave paternity issues aside, the believed degree of biological closeness does not need to correspond closely to the proportion of genes actually shared. For example, there are many people who believe themselves to be more closely related to parallel cousins than to cross-cousins, although they share the same proportion of genes with both categories.

Part III
Encounters with Modernity, Sudan and Sudan–Ethiopia Borderlands

Chapter 6

The Rise and Decline of Lorry Driving in the Fallata Migrant Community of Maiurno on the Blue Nile

Al-Amin Abu-Manga

Maiurno on the Blue Nile in central Sudan was founded as the last station of the Fallata hegira (religious migration) from the defunct Sokoto Caliphate (northern Nigeria) after its fall into the hands of the British colonial army in 1903. Its name derives from that of the Fulani Sultan, Mai-Wurno, who first settled there with his people and followers. The first generation of settlers remained conservative, resistant to any influence of the host communities and suspicious of anything connected to the colonial administrators, whom they considered to be 'unbelievers'. But the immediate descendants of the first migrants proved to be more enthusiastic – even more so than the local people – towards the instruments of the modern colonial economy in Sudan. Lorries were one of the first and most important of these instruments, and the lorry drivers – those who steered these instruments – obtained a novel and remarkable status from their role. Lorry driving started to attract recruits from among the Maiurno youths from the 1930s, and lorry driving reached its zenith in the 1960s and 1970s. But in the last fifteen years this profession has witnessed a drastic setback and decline.

My chapter traces the introduction, development and decline of lorry driving in Maiurno.[1] Through this, I examine the perceptions that the Maiurno people had of themselves, their place in the world and the changes that have taken place from this early settlement until today. Unlike Duffield (1981), who treated this subject from a capitalist point of view, our study focuses on the glorious image of the lorry driver as reflected in the eyes of the Fallata in Maiurno and the other Fallata-related communities along the Blue Nile. It will argue that the fantastic image that

1. This chapter is based on data that have been collected since 1996 by Prof. G. Schlee and the author on the Fallata communities of the Blue Nile within the framework of the project: 'Ethnicity in a new context'. The author is indebted in this regard to the German sponsoring bodies (DFG and Max Planck Institute) as well as to Prof. G. Schlee, the initiator and head of the project. The author is also grateful to Dr Leoma Gilley for comments on the wording of the first draft of this chapter.

developed derived not only from the financial comfort of the lorry driver; there were
other more important factors that contributed to its creation. The decline of the
transport business in Southern Blue Nile over the last fifteen years has negatively
affected the lorry driver's financial comfort, but his image started to lose its glitter
some years before. This study also tries to examine why this is the case.

The Formative Period of Maiurno Society

Present-day Maiurno town lies on the western bank of the Blue Nile, 15 km south of
Sennar, and covers an area of approximately 8 sq. km. It has c. 30,000 inhabitants. The
major ethnic groups living in it are the Fulani (71.7 per cent), Hausa (16.4 per cent)
and Songhai (7.7 per cent).[2] Arabic is widely spoken by almost everybody, but language
use in Maiurno is also characterized by multilingualism (Fulfulde, Hausa and Arabic).

All informants emphasized the simplicity of life in Maiurno during the first three
decades after its establishment (1900–30) and yet evidence from our data and archival
sources show that this new settlement witnessed its most turbulent life during this
period. Many new waves of migrants from West Africa and from other regions in
Sudan followed the steps of the hegira and joined Sultan Mai-Wurno in Maiurno.
Some clan leaders went out of Maiurno to found their own settlements with the
consent of Mai-Wurno,[3] while others, namely the core of the Borno Fulani, left in
resentment because of mistreatment by the Maiurno ruling circle.[4] Other groups
headed southward along the Blue Nile and eastward to River Atbara and River Setit to
escape from famine.[5] This period also witnessed the construction of the Sennar dam,
for which most of the physically able Maiurno people provided labour. There were also,
at this time, internal disputes among those competing for power. Mai-Wurno came
into conflict with the colonial administration because their policies were unsettled and
they were unclear about what position and recognition they were to award him.[6] Other

---

2. The bulk of these are originally Fulani who were assimilated into Songhai culture and lost the
   use of their ancestral language (Fulfulde) long before their migration to Sudan. The
   percentages indicated above were obtained from a language survey carried out by the author
   in 1977 while preparing for his MA.
3. These include the Katsina Fulani richest migrant, Faruku, who founded Hillat al Beer (jointly
   with a Borno Fulani clan leader, Jawro Hamma), Dindir Kawli and Wad al Obeid
   (information from Omar Jum'a Muhammad, eighty-four years old, born in Dindir,
   interviewed in Maiurno on 08 November 2000). They also included Mai-Laga, a Sullubawa
   Fulani clan leader, who left around 1925 to found the village of Mai-Laga on River Setit in
   eastern Sudan (information from Osman Hassan Muhammad, eighty-three years old, born in
   Maiurno, interviewed by the author and Catherine Miller in Mai-Laga on 08 March 1996).
4. The best example is that of Sultan Ahmed of Misau, who returned from Mecca in response to
   his people's request and moved away with them to establish his own settlements (Galgani and
   Abd al Khallag, c. 70 km and 75 km, respectively, south of Maiurno on the Blue Nile) in 1915.
5. For example, the founders of the villages of Basis Hamma Kumbo and Wuro Jibo, both on the
   western bank of the Blue Nile, about 80 km and 85 km, respectively, south of Maiurno.
6. In 1930 Mai-Wurno burst out in front of the DC of Mukwar: 'I am not a fool. I know why
   the Government treats me with undeserved suspicion thinking that I will stir up trouble',
   Sudan Central Records (SCR): 36-D-15, 4 February 1930.

Maiurno people were also opposed to the colonisers' strong promotion of secular schooling.[7]

It was under these conditions that aspects of modernity started to take shape in the village, mostly coming from outside. The first non-local (Arab) merchants came from the Gezira, from al Jadid, some 30 km south of Khartoum. Simultaneously, a group of Hausa traders emerged, notably Abarshi and Mai-Saje. When the first group of tailors[8] arrived in Maiurno in the mid-1920s, the market was, by and large, dominated by the Hausa traders. As sewing machines were installed in shops, the relatives of these Hausa traders monopolized the early opportunities to be engaged in this new 'ultra-modern' profession.[9] In 1925, Sultan Mai-Wurno's compound moved away from the river to its present site, while Dan Galadima Hassan and his brother Wali, the Sultan's successive viziers, were left behind as shaikhs of the Hausa (traders') quarter. It is worth noting that the members of the Maiurno ruling family, although originally Fulani, spoke Hausa as their first language at that time, and were thus identified as Hausa rather than Fulani. So the result of the above association, as noted by Duffield (1981: 37), 'was to render Hausa the language of both market and authority'.[10] This, of course, had some bearing on Fulani/Hausa relations in Maiurno. It created a feeling of dissatisfaction among the Fulani, who constituted the great majority of the inhabitants, leading some of their clans (mostly the Borno Fulani) to migrate from the village, as noted above.

However, this situation did not last for long. The emergence of lorry driving in the 1930s changed the balance of power. The first three generations of Fulani engaged in this occupation balanced out the prestigious comfort enjoyed hitherto by the Hausa group of merchants. Later on, it also overturned the situation in terms of affluence and prestige.

Introduction of the Lorry and Lorry Driving in Maiurno

The first lorry in Maiurno was bought by Sultan Mai-Wurno in 1930. This event was associated with a series of disputes between Mai-Wurno and the colonial administration, engendered indirectly by the way Mai-Wurno wanted to secure money for this purpose. By 1928 the Sultan had already established two of his wives in Dindir Kawli, a village founded by his followers on River Dindir (approximately 35 km east of Maiurno). His intention was to make Dindir Kawli the second seat of

---

7. For the reasons for, and ways of resistance to the secular school by the early Maiurno generations, see Abu-Manga (1993).

8. Some of them were Arabs from Gezira, whereas two of them were Borno (Kanuri) from Gedaref.

9. The first two local tailors were Abbakar Dabalo and his brother Tahir, and the latter was the first Maiurno citizen to own a sewing machine. However, both of them later shifted to trade and opened shops in the market, thus adding to the group of Hausa merchants.

10. Now the situation has changed considerably. Today Fulfulde is spoken in the Sultans' compound more than Hausa, and the family has started once again to be regarded as Fulani, which suggests that language is stronger as an indication of identity than ethnicity. More discussion on this point will be made in a subsequent work.

his court for ruling the newly founded Fallata settlements in that area.[11] However, he seemed not to have needed the lorry only 'for visits of his villages in Dindir and for transporting his grains', as established by Nasr (1980: 11–12) and Duffield (1981: 115), but probably also to compete with the other heads and chiefs enrolled in the Native Administration, in terms of status and prestige.

Up until this time he had no source of capital, so he dispatched his son Muhammad Tahir, in December 1929, to collect money from the Fallata settlements established by his followers in eastern Sudan (Gedaref/Kassala area). But by that time a new Native Administration law had already been promulgated which gave him no rights to these settlements. Accordingly, Muhammad Tahir was ordered to return from Gedaref, an action that stirred the fury of Mai-Wurno against the colonial administration. After a series of letters between Kassala, Gedaref, Medani and Makwar (now Sennar), in April 1930 Muhammad Tahir was given restricted permission to visit Gedaref and Kassala towns only (and not the surrounding Fallata villages).[12] However, even in these towns he was able to collect 'abundant money – sheede kifaaya', as stated by an informant[13] and confirmed by colonial reports: 'His son had been to Gedaref and Kassala; he had been able to scrape together enough money to purchase the motor lorry which he had set his heart on, and to all outward appearance he regarded the incident of his son's treatment at Gedaref as closed.'[14]

The story of the first driver of that lorry – Awad Salim by name – is not shorter than the lorry itself. Awad Salim, identified as Dongolese by one informant and as Malakiyya (ex-slave) by another, was working for the company where the lorry was bought. He was asked to drive the lorry to Maiurno and bring back the number plates. But Awad Salim then decided to continue working in the lorry, although Muhammad Tahir had already promised the job to somebody in Kassala. At last Awad Salim settled permanently in Maiurno, married one Fulani woman and then another Mawalid Fulani woman originally from Shaikh Talha. He gave birth to many children, who grew up as Maiurno citizens with a rich linguistic repertoire (speaking Fulfulde, Hausa and Arabic).[15]

---

11. Maybe following the tradition of some of his predecessors in the Sokoto Caliphate, notably Muhammad Bello (after whom he was named), who used to have court seats in both Sokoto and Wurno in present-day northern Nigeria.
12. See the details on this correspondence in: From DC Gedaref to DC Wad Medani, 18 December 1929, SCR (unclassified); BNP/B/36.G.3, 9 January 1930; KP/SCR/35.A.2, 20 January 1930; SCR, 36-D-15, 4 February 1930; BNP/SCR/36.G.3, 13 February 1930; KP/SCR/35-A.2, 18 March 1930; BNP/SCR/36.G.3, 21 April 1930.
13. Omar Jum'a Muhammad (see note 3).
14. From Governor, Blue Nile Province to the Civil Secretary, Sudan Government, Khartoum, BNP/SCR/36.G.3, 21 April 1930. As an aftermath of this 'lorry incident' the colonial government decided at last to demarcate a strip of land for Mai-Wurno and to give him some kind of title or status within the framework of the Native Administration.
15. In the 1960s he moved to Sennar and died there. His Fulani widow left afterwards for Saudi Arabia to join her former husband in Medina. When the author performed the pilgrimage in 1995, he found her still physically and mentally fit. She asked about many Maiurno people, most of whom were already dead. She died two years ago, aged over 100.

A few months after the purchase of the first lorry, the Sultan brought another, second-hand one, this time not out of necessity but in order to fulfil the promise made by his son to the Hausa driver from Kassala – Ibrahim Bayaza by name. That person, according to one informant,[16] used to have two lorries of his own. When Muhammad Tahir, who happened to be travelling in Ibrahim Bayaza's lorry, expressed his wish to find a driver of his calibre for the lorry his father intended to buy, the man got rid of his own lorries and came to Maiurno to serve Dan Fodio's descendants in order to gain their blessings. But, as mentioned above, 'the Arab refused to step down'. The Sultan was then obliged to buy another lorry so as to create a job for the Hausa driver. Towards the end of the 1930s, a third lorry was bought jointly by the Sultan and a merchant of Syrian origin coming from Wad Medani.[17] In order to be well established in Maiurno, that merchant married into Mai-Wurno's family and opened a shop in the market. During the 1940s and 1950s, lorry ownership became independent from the monopoly of the royal family and extended to local merchants and successful lorry drivers, who, with the exception of one, were all Fulani.

With regard to lorry driving, our informants disagree on where the first generation of Maiurno drivers received their initial training in this profession. While one informant[18] insists that they all started and were trained outside Maiurno (Medani, Khartoum, western Sudan), another informant asserts that they started their apprenticeships in Maiurno under the first three outsiders and then, later, travelled outside, where they received advanced training until they became drivers.[19] Unfortunately, five of the six members[20] of this first generation are now dead, and only the surviving member (Muhammad Ibrahim Sajo, eighty-seven years old) could be interviewed. He emphasized with a kind of pride that he was trained by the *khawaajaat* (white men),[21] which may support the assertion of the first informant. However, it is important to note that the 'Italian War' (the Second World War)[22] constituted a significant factor in their career, not only in terms of the skills and experiences gained during their service in Eritrea and Ethiopia, but also regarding the orientation of their characters and behaviour, as will be seen later. But, with all their adventures in the above war, only one member of this generation was successful enough to own a lorry.

Most of the members of the second generation of Maiurno drivers received training under Arab drivers in Wad Medani, Khartoum and western Sudan.[23] They

---

16. Omar Jum'a Muhammad (see note 3).
17. Duffield (1981: 115) listed it erroneously as the second.
18. Omar Jum'a Muhammad (see note 3).
19. Abd al Rahman Ali Babikir, sixty-three years old, born in Maiurno, interviewed in Wad Medani on 09 December 1996.
20. These were Bazagalli, Abbakar Busa, Muhammad Sajo, Muhammad Bello Abd al Wahab, Yahia Galadima and Ahmad Duddu.
21. Interviewed in Maiurno on 13 November 2000. His master was called Michael, pronounced Misheel, which sounds like a Lebanese Christian. M. Sajo was also proud of having been examined for the driving licence by a British man. His driving licence is dated 1935.
22. In reference to the Sudanese campaign in Eritrea and Ethiopia during the Second World War.
23. They include people like Osman Shanuwa, Hamid Parapato, Mu'az Abdullahi, Abbakar Mai-Katuru, Sanda Sawamil, Siddig Saksaka, Muhammad Abbakar Moli and Shindo Turaki.

then came back and started to operate regular trips between Sennar/Singa and southern Blue Nile to Rosseiris, Kurmuk, Geissan and even inside the Ethiopian territories, some as hired drivers and four of them with their own lorries.

All our informants agreed that lorry driving in Maiurno reached the apex of its glory in the 1960s with the emergence of the third generation of drivers whose members were far more numerous than those of the preceding generations.[24] Until the last years of the 1960s, only two or three members of this generation had lorries of their own; all the others worked as hired drivers, based mainly in Singa. At that time Singa was the central market for gum arabic and sesame, produced in the areas of Dali and Mazmoum.

Many factors contributed to the ascension of lorry drivers to a distinguished class in Maiurno in this specific period. Members of all the three generations overlapped and associated together to constitute a large occupational group. Secondly, the transport business in Sudan at that time flourished considerably, and this eventually reflected positively on the drivers' income. So, with their high income, they had access to all means of comfort that were inaccessible to the other sectors of the Maiurno society. Thus they soon came to surpass in status the hitherto prestigious classes of the merchants and tailors. This situation continued as such for more than two decades. When Duffield was doing his research in 1977 there were forty-seven lorries owned by Maiurno people: six owners had two lorries and one had three (Duffield 1981: 113). The number of drivers in that year reached 160, the overwhelming majority of whom were Fulani, which led Duffield to conclude that 'lorry driving and especially lorry ownership is largely the prerogative of the descendants of Maiurno's Fulani and not its Hausa settlers' (ibid.). So, if in the 1920s Hausa was 'the language of both market and authority', in the 1970s Fulfulde became the language of lorry driving and the transportation business, and it started to displace Hausa in the 'market' and 'authority' domains.

The Glorious Image of the Lorry Driver in Maiurno and the Related Fallata Settlements

An informant recalls: 'We started to see lorries here in Maiurno since the time when the "unbelievers" [probably Greeks] used to pass through on the way. We usually went out, with nothing on us except our pants, to watch the lorries'.[25] So the first sight of lorries in Maiurno was associated with the 'unbelievers', the 'business' of whom the first conservative generation of the migrants resisted strongly, including their (i.e. the 'unbelievers') secular system of education and their way of dressing. It is paradoxical that, in the end, Maiurno became indebted for its social and economic development to this 'unbelievers' thing'.

---

24. Included here are people like Musa Hassan, Suleiman Abu-Musba, Abd al Gadir Jemis, Ahmad Omar, Abbakar Suleiman, Abbakar Muhammad Bello, Abdullahi Jabir and Abu-Manga Muhammad.
25. Muhammad Abbakar Moli, seventy-eight years old, born in Maiurno, member of the second generation of drivers, interviewed in Maiurno on 14 November 2000.

We have seen that all the six members of the first generation of drivers served in Eritrea and/or Ethiopia in the early years of their occupational career during the 'Italian War'. Some of these drivers seemed to have been outstanding on the war front, which earned them a praise song by the famous woman singer, the late Ai'sha al Fallatiyya.[26] An informant[27] recalls some verses of this song, whose words run as follows:

> Return once again; return once again
> Gondar[28] is far, oh usta ... Allah
> He pressed the accelerator until it rang
> He knocked the gear shift lever until it sang ... Allah
> Don't let a child overcome you
> Don't let the hills overturn you ... Allah
> ... its [the lorry] driver is from the *kai wannan*.[29]

Such folk literature about lorry drivers is abundant in the Fallata settlements along the Blue Nile south of Maiurno. One of the popular songs in that area (and in Maiurno as well) is *Gunduwaare*, which eulogizes a number of the first and second generations of the Fallata lorry drivers.[30] An informant[31] relates that during the 1950s, when a driver was to stay overnight in a certain village in that area, girls would bring him fried chicken and boiled milk and spend the night chatting (innocently) with him. He reported that, when he walked outside, married (secluded) women would make holes in the straw or stalk walls (mats) of their compounds to watch him with admiration. There was even a woman in that village who was so infatuated with lorry drivers that she sometimes accompanied some of them on their trips. Her nickname appears in the above-mentioned popular song of *Gunduwaare*. Other descriptions of the lorry driver of that time included in our data are the following: 'he was more important than a government employee'; 'he was the *mughtarib*[32] of

---

26. She was the first Sudanese woman singer heard on the radio. She was of Hausa background, taken to the front to entertain the Sudanese army fighting in Ethiopia during the Second World War.

27. Abd al Rahman Ali Babikir (see note 19).

28. A town in Ethiopia near the Sudanese border, one of the fighting fronts.

29. A Hausa expression meaning 'oh you'. In the Sudan it symbolizes the Fallata in general (i.e. all people originating from West Africa). A colleague at Khartoum University coming from Kassala (near the Eritrean border) affirms that this specific song is well known in his hometown.

30. Such folk literature, however, might not be confined to the Fallata settlements, because an epithet in praise of an Arab village called al Dali (on the way between Singa and Mazmum) exists in Arabic, probably composed by its inhabitants. It says:

Al Daali habiibak	Al Daali is your beloved (village)
Abu-Manga yiwaddiik	Abu-Manga takes you (to it)
Wa l khaawaaja yijiibak	And the white man brings you back

31. Omar Jum'a Muhammad (see note 3).

32. One working in a Gulf oil country. The *mughtarbiin* are associated with abundance in every aspect of comfort.

that time'; 'he was compared to the head of state'; 'he was higher in the eyes of the public than the president of the world'; 'he was a wonder, he was one of the wonders'.

All our informants agreed that the driver derived his greatness primarily from his ability to drive the lorry and not from his comfortable financial position. This is also evident from the fact that he was held in higher esteem than the lorry owner (i.e. his employer). 'This driving used to be [regarded as] a wonderful thing,' says an informant. 'When he [i.e. the driver] came out, he would walk swaggering.' At that time the drivers used to wear short trousers, identified with the 'unbelievers' and government employees. Even children playing lorry games used to adopt the names of distinguished drivers.

The position of the drivers in the Fallata society of the Blue Nile combined with their magnificent image in this society contributed effectively to the psychological formation of their personalities, developing in some of them a complex of greatness and superiority. Some of them came to be known for inconsistent and contradictory behaviour. For example, one of the drivers, having worked for some time under the Italians in Ethiopia during the Second World War, strove to behave in a European manner. He wore a European mode of dress, even at a late age, which was unusual in Maiurno society at that time. He also used to break wind in public, an act that is considered very shameful in the Fulani culture. He did that – he said – because the Europeans did not find it shameful.[33] He usually did not greet people whom he did not know very well, and he might not respond to others' greetings, especially early in the morning. Another driver[34] boasts of being trained by the *khawaajat* (white men). In the interview he emphasized that, at the time he became a qualified driver, *'an naas ma btaʿrif haaja'* – 'people didn't know anything'. The use of the third person feminine (*btaʿrif*) in reference to 'people' in this phrase has pejorative connotations in Sudanese colloquial Arabic. This person had hardly any friends in Maiurno, even among his fellow drivers. He used to be tough with the passengers; he did not hesitate to insult a passenger or spit in his face or even ask him to step down from his lorry in the middle of a trip. Another driver used to smoke marijuana in public, though it is not only socially unacceptable but also prohibited by law.[35]

These examples should not be taken as general characteristics of the old Maiurno drivers. They are some extreme cases confined to individual members of the first and second generations. The anomalous behaviour described above was engendered by the large gap that existed between their experiences in the urban milieu where they spent the first years of their career (including their travels and adventures during the 'Italian War'), on the one hand, and the simple and conservative society (of Maiurno) in which they later came to live, on the other. At the other end of the continuum of behaviour and style there were many drivers with normal and even modest characters, with the majority somewhere in between.

---

33. He once advised the author to break wind whenever he felt like it – for the sake of his health.
34. Interviewed in Maiurno on 13 November 2000.
35. This same person, being a colleague of the author's father, smoked marijuana many times in the author's presence while the latter was a child.

Development of the Maiurno Drivers into a Distinguished Class

We have seen that in 1977 there were about 160 drivers in Maiurno. Of course, these did not all have the same status. They could be classified into first, second and third class drivers according to a number of variables, such as age, experience, skill, lorry ownership and other personal qualities. A group of five or six junior drivers might have been trained under a single senior driver. Therefore, another possible classification of these 160 drivers would be to group them according to masters (usually few in number) and students (usually many in number).

By the time Duffield was doing his research in Maiurno (1977–79) the lorry drivers had already formed a distinguished class. The emergence of this class followed a gradual process that started to take shape in the early 1960s, when the number of the drivers ranged between thirty-five and fifty. At that time there were no asphalt roads in the entire country. During the dry season the majority of Maiurno lorry owners and drivers were to be found south of Sennar in the areas of Mazmoum, Rosseiris, Geissan and Kurmuk, which were the main commercial centres in Southern Blue Nile region. By the end of May, when the rains were about to begin, their activities there would start to lose momentum, and, from the middle of June until October, the roads south of Singa were virtually closed to lorry traffic. Almost all the lorry owners and drivers would return to Maiurno, some hired drivers coming along with their lorries and some others parking them in their owners' respective hometowns or villages. So 'at the end of the season' (in their words), the hired drivers would collect all their savings and return to enjoy a pleasant three-month vacation. This vacation used to be a peak period, not only for the drivers but also for the village in general. Their excessive expenditure used to animate the economic life and their handsome appearance used to attract the admiration of people, especially of children and young girls. For 'not only do drivers form friendship groups amongst themselves, but they strive to maintain the appearance of sophistication and affluence with their fluent Arabic, clean, pressed gowns, dark glasses and heavy wristwatches' (Duffield 1981: 113). Some drivers were known for their particular styles of turban winding. Every well-established driver had to have a blanket, the value of which matched his status. For example, first-class drivers possessed expensive wool blankets. At a certain time they also possessed coats for winter.

As mentioned above, during their vacation the drivers associated amongst themselves in the form of friendship groups, usually according to their generation. They used to take breakfast (around 9.00–10.00 a.m.) in restaurants. By that time they had already bought foodstuff for lunch (lamb meat, fish, chicken, vegetables, salad, etc.) and sent it to their respective homes. Before midday, members of each group would gather in the house of the colleague who had a convenient guest-room to play cards.[36] The owner of the house sometimes offered lunch to the entire gathering; otherwise a number of them would send for food to be fetched from their homes and to be eaten collectively. They would disperse in the early evening. Some of them would meet once again in the late evening to chat, while others reserved this time for the family.[37]

---

36. The house of the author's father used to be one of the gathering places.
37. The author does not remember seeing them taking alcohol.

Until the 1980s, the Maiurno drivers did not have a formal organization and yet they were 'united as one block', according to one informant. They used to exchange visits with the Singa (Fallata) drivers, whereby reciprocal grandiose hospitality was offered, including the slaughtering of rams. Before the end of their vacation, some lorry owners would come to Maiurno to book drivers. The Maiurno drivers were highly valued by Arab lorry owners, because of their integrity, loyalty to their profession and hard-working nature: '*Sawwaagiin Maayrino 'indahum sum'a 'aalamiyya* (The Maiurno drivers have international fame [i.e. very high value]),' states an informant.[38] 'If you didn't hire a driver from Maiurno, [it was as if] you hadn't done anything.'

This pattern of the drivers' life and socialization during their vacation continued to varying degrees until the early 1980s. The flourishing of the transportation business in Sudan during that period largely facilitated their lavish lifestyle.

During the height of their social and economic position, lorry drivers were the dream of every girl of marriageable age. A number of lorry drivers married into the Arabized Fulani and Arab families of Shaikh Talha, from whom wives were, otherwise, not easily accessible to the simple (vernacular-speaking) Fallata of Maiurno.[39] It was during this period that rich Maiurno people of all occupational and professional classes, especially successful lorry drivers, came to own the highest number of efficient lorries.[40] At that time (and until recently) 'lorry ownership is seen by the bulk of Maiurno's population as the zenith of social and economic success' (Duffield 1981: 116). Lorry drivers were among the first Maiurno citizens to construct relatively beautiful houses and get access to electricity and piped water.[41] Some of them were enlightened enough, through contact with the Arab communities (of their employers), to be able to realize the importance of Western (secular) education and thus to secure it for their children.[42] Their wives used to compete in wearing the most fashionable clothes. The drivers' travels and visits to big towns and cities, in addition to their financial affordability, permitted them to acquire items of fashion and modern gadgets very early on, such as beautiful clothes for themselves and their families, radios, cassette recorders, etc.

---

38. Abu-Manga Muhammad (see note 24).
39. Shaikh Talha is situated across the river. Its inhabitants were a mixture of old immigrant Arabized Fulani and many other Sudanese (Arab) tribes. Even the originally Fulani among them identify themselves as Arabs, and thus regard themselves as being superior to the vernacular-speaking Fallata of Maiurno. So, in many cases acceptance of a Maiurno husband by a Shaikh Talha family indicates the social and financial competence of the former. Maiurno-Shaikh Talha relations will be the subject of a subsequent paper.
40. The author's father owned two efficient lorries, a tractor and a Land-Rover for private use. See History A in Duffield (1981: 116).
41. Muhammad Ibrahim Sajo (see note 20), a member of the first generation of drivers, was the third person in Maiurno to build a house of brick – after the Sultan and A. Kagu. He was the third to join electric power to his house. Interviewed in Maiurno on 13 November 2000.
42. All children of the author's father attended the school. His firstborn is the first in the area to reach the status of 'Professor', and one of his daughters is the first female in the area to reach the status of 'Assistant Professor'.

From the mid-1960s to the mid-1980s the lorry drivers constituted a distinguished social class in Maiurno. We described them as a class because they were 'united as one block', and they had a distinctive outer appearance. In addition, they had their own jargon, derived mostly from the lorry parts (moving around like the fan, standing still like the brake, the woman's differential (pronounced *difirinshi*) in reference to her (bulky) lower part). Due to their close contact with the Arab societies of their bosses, many of them developed the habit of swearing 'by divorce' (*ᶜalay at talaag*),[43] which is unknown in Fulani culture. But, above all, they had conventional ethics and values that they observed as strictly as possible.

The Conventional Ethics and Values of Lorry Drivers

The Maiurno drivers' society used to be distinguished by a series of ethics and values, some of which were also shared by drivers of other non-Fallata societies. Some of these ethics and values were dictated by the rough topography and the strangeness of the people and geography of the regions in which they ran their activities. The ethics and values provided by our informants are summarized in the following points:

- Maximum assistance to a colleague whose lorry had broken down in the bush. This included waiting and helping in the repair of the fault, offering minor spare parts or lending the expensive ones (such as tyres, etc.). This is in addition to offering food and water to the lorry assistant guarding a broken lorry.
- A driver had to be honest and trustworthy vis-à-vis the lorry owner. If he was not, he would lose the sympathy of his colleagues for having damaged their image.
- A jobless driver was not expected to have his colleague sacked in order to replace him (an act referred to in their jargon as *ᶜifriita* (jacking)).[44] If a driver was sacked by a lorry owner after a long period of service, his colleagues were not supposed to accept employment with that lorry owner.
- Financial assistance to a driver who had lost his job, including invitations for breakfast by his colleagues and provision with pocket money (known as 'cigarette money', even if he did not smoke).
- A driver travelling in a lorry driven by a colleague would not be charged any fare. He would, in addition, be fed by his colleague throughout the journey. If the lorry owner was the one managing the lorry and he insisted on charging the guest, the host driver would ask the owner to deduct the cost from his salary; otherwise he would threaten to 'step down' (leave his job).
- When a group of drivers met (accidentally) in a restaurant, the first one to get up would pay for all the others. Usually everyone would try to precede the others.

Lorry Owner-Driver Relations

Most of the lorries driven by the Maiurno drivers belonged to Arab merchants in the Gezira (between Medani and Khartoum), Sennar and Singa. A few of them belonged

---

43. For example, 'If it is not so, I will divorce my wife.'
44. A jack is a portable device for raising the axle of a motor vehicle so that a wheel may be changed.

to merchants or successful former drivers from Maiurno itself. When a lorry owner
was to engage a driver, this was usually done through verbal agreement (as to the
salary) between the two parties. In other words, there was no question of a written
and signed contract, since the majority of the drivers and many of the lorry owners
were illiterate. No witnesses were even required, because the relation between the two
parties was that of absolute mutual trust. All the driver needed was a notebook in
which his deductions during the season were to be recorded. Apart from his salary,
he was also paid a certain amount of daily pocket money, even if the lorry, for any
reason, was not travelling for many days. The driver was entrusted with the lorry and
would usually be advised by the owner that 'money is not important; just keep the
lorry in a good condition'. He might sometimes be left to travel about to seek for
loads in his own way and bring the revenue. In other cases the owner would load the
lorry with general supplies and merchandise, such as tea, sugar, oil, soap, salt and
bales of cloth, which the driver would sell on his southward trip to the Southern Blue
Nile. From there he would bring hardwoods, roofing laths, palm fronds and other
material for furniture and building with the money he made on the way. These two
systems of working conditions were ideal for the driver, because during the entire
journey he had an open fund for his financial needs and hospitality, and had more
freedom to fulfil his obligations towards his colleagues, which would keep him in
high esteem. This would also give him an opportunity to prove his integrity. An
informant[45] reported that he had been working for a merchant in Wad Medani for
many years under this system without having a single case of mismanagement or any
doubt as to his integrity by the lorry owner.

A third system of working conditions, which was deemed unpleasant for the
driver, was the case in which the owner himself or someone delegated by him
(Arabic, *wakiil*) carried out the task of managing the lorry, thus reducing the driver
to a mere 'steering-wheel driver'. Not only was his freedom regarding hospitality
expenses and the satisfaction of other lorry-driving ethics and values considerably
curtailed, but also it would annoy him that the owner or his *wakiil* would keep
intervening in matters related to the way he was driving. Therefore, drivers, especially
of the later generations, started to refer to *wakiils* (including the owners following
their lorries) by terms associated with negative connotations such as *ᶜaarid* (obstacle),
kataawit (cut-out) or *mufattish* (controller). However, it seems that there were also
cases – though very rare – where the driver and the lorry owner (or his *wakiil*)
worked harmoniously and remained friendly with each other.[46]

What the lorry owner did not like most on the part of the driver was to see his
lorry and the driver always dirty or the lorry often broken down. 'You were a good
driver if you travelled from here [Maiurno] to Kurmuk and back without breaking a
spring sheet,' confirmed an informant. Such a driver would be regarded as a model.

---

45. Ahmad Omar Muhammad, sixty-five years old, born in Maiurno, interviewed on 19
    December 2001.
46. Abu-Manga Muhammad (see note 24) quotes himself as one of these cases. At a certain time
    his employer's nephew followed him as *wakiil* and the two of them were on good terms: 'He
    was like my son.'

On the other hand, apart from the *wakiil,* the driver did not like to see the lorry owner seeking information about his conduct, asking if the driver had the habit of taking alcohol or visiting prostitutes in other words 'spying on him'.

In the case of a disagreement between the owner and the driver before the end of the season, the latter would 'step down' (leave his job) by one of the following three types of symbolic acts: (1) carrying away his blanket; (2) throwing the ignition key to the owner; or (3) asking the owner to 'give him his account' (i.e. to calculate his deductions and give him the rest of his entitlement for the period of his service).

First-class drivers were usually known for their stability and long service in the same lorry, due to their unquestionable integrity, good experience in management, seriousness and charisma. In fact, it was these attributes that qualified them as first-class drivers. Such drivers usually continued smoothly up to the end of the season, i.e. the beginning of the rains. When leaving for their vacation they would get all their rights from the lorry owner amicably.

In the preceding sections we have tried to describe the position of the lorry driver in Maiurno during 'his days', right from the beginning up to the dawn of his heyday in the mid-1980s: his actual position, the picture he strove to portray of himself and his fantastic image in the eyes of the society. Throughout our description we used the past tense, because this situation has now changed drastically, and what we have said above no longer applies to the present-day lorry driver. But, before discussing the various factors behind this change, it is necessary to explore the nature of the connection of the Maiurno lorry drivers with the Southern Blue Nile region, where they mainly ran their occupational and commercial businesses.

Maiurno Drivers in the Southern Blue Nile Region

The Southern Blue Nile (Arabic, *As Saʿiid* = the South) plays an important role in the history of lorry driving among the Fallata community of Maiurno and its related settlements. This region geographically comprises the area extending roughly from Wad al Nayyal south to the Ethiopian border and the northern edges of the (former) Upper Nile Province. The major commercial and/or administrative centres of this region are Rosseiris, Damazin, Geissan, Kurmuk, Mazmoum and Bunj. Almost all the Maiurno drivers were at one time connected with this area, either as lorry assistants, drivers or businessmen (after having retired from driving). Eight of them got married there,[47] and half of these succeeded in establishing stable families, in addition to their Maiurno families.

To understand the nature of relations, connections and interaction between the Maiurno drivers and the different categories of people living in this region, we first need to delimit these categories as follows:

---

47. These are Abbakar Suleiman (nicknamed Kharrim), Abbakar Muhammad Bello (nicknamed Mastuul), Siddig Saksaka, Al Zubeir Abdalla Ali (nicknamed Basha), Abd al Rahman Ali Babikir, Abd al Halim Mahmoud, Ibrahim Abu-Bakr (nicknamed Iro) and Ali Abbakar Busa.

1. Local inhabitants of the region (Berta speakers, Gumuz speakers, Hamaj, Ingessana, Uduk, Burun, etc.), including the Fallata immigrants (Fulani, Hausa and Borno (Kanuri)).
2. Jallaba Arab merchants who have not yet been integrated into the local communities.
3. Government employees, mostly policemen, teachers, medical assistants and (male) nurses, and recently soldiers and army officers. With the exception of the soldiers, these employees have mostly come from the Arabs (or Arabized groups) from both inside and outside the region.
4. Lorry drivers from different ethnic groups, with the Fallata constituting the largest single ethnic group.

The place of the lorry driver in the socio-economic network involving the above categories of people is dictated and explained by another important factor, namely, the underdeveloped nature of the area with regard to the fundamental services (schools, hospitals, etc.), especially roads. The only existing asphalt road ends in Damazin. As this region falls within the zone of heavy rainfall, all the roads from Damazin in every direction except the north remain closed to traffic for at least four months of the year. The topography of the region, with its many streams and hills, is considered harsh by the non-local inhabitants. Thus, the Jallaba Arab merchants, the government employees and the drivers share the common factor of 'living in a strange land', and that unites them and creates in them a feeling of solidarity in opposition to the local inhabitants of the region. This solidarity is usually expressed and affirmed through a mutual exchange of free services and gifts.

The lorry driver, however, is the central figure that links the other categories of people together and connects them with the outside world: he brings merchandise from the northern towns to the Jallaba Arab merchants stationed in that region and transports the region's products for sale in the north. In the course of his travels he also carries messages and variable goods back and forth between the Jallaba and their respective families in the Gezira. He transports the government employees to their posts in the remote areas and brings them their needs from the towns. He may also carry messages to their respective families in the north. Until recently, he used to transport the schoolboys and students from the remote villages of the region to their boarding houses in towns and bring them their necessities from their parents.

Evidence from our data indicates that the Maiurno drivers are the closest outsiders to the local inhabitants of the area. They are the people who were able to penetrate into the remote small villages carrying important items such as sugar, soap and clothes to the villagers, in addition to 'aspects of civilization'. Many of the Maiurno drivers are known for having opened roads to certain areas.[48] But, above all, for a long time they have been providing the villagers with the main opportunity for

48. E.g. Faroug al Digel opened a road in the area of Fadamiyya (eastern side of the river); Abbakar Kharrim and Abbakar M. Bello opened many roads in the area of Yabus; Abu-Manga's lorries opened roads in the areas of Malkan and Samᶜa.

labour and income. Until the escalation of the civil war in this area in the late 1980s, the main and – for some people – only cash products of the area were bamboo, palm fronds and hardwood. Growing these items does not involve any cost, because they are found naturally in the bush. The inhabitants cut or collected them for sale to the drivers, or the drivers hired labourers from the local people to do the job. Maiurno drivers such as Abbakar Kharrim, Abbakar Muhammad Bello and Zubeir Basha were able to broaden their businesses in this way. The three of them established families there and became more identified with that area than with Maiurno. They all confirmed that they did not encounter any communication problems, as Arabic is widely used as a lingua franca. Yet, through long and close contact with the local people, they were able to pick up some words and phrases of Berta and Uduk. However, the rest of the 100–150 Maiurno drivers centred in Damazin before the war were not on such close terms with the local inhabitants.

The image of the driver in that area was greater even than it was in the north, owing to the role he played in the general socio-economic setting of that region. His business was not confined to simple transportation, as he also secured labour and income for the local people. Besides that, he assumed the role of a postman and was regarded as an explorer, adventurer and messenger of civilization. Until recently the first driver (or convoy of drivers) to reach the distant towns of the regions (e.g. Kurmuk and Geissan) after the rainy season was (were) received with cheers and ululation by a large number of the citizens (of both sexes) outside the towns. For many months he (they) would be seen in the eyes of the citizens as the hero(es) of the year.

The civil war in the extreme southern, south-eastern and south-western parts of the region redivided all the inhabitants into just two groups: the rebels and the victims, irrespective of any ethnic considerations. During the assaults on Kurmuk, Yabus, Chali and Bunj, the Maiurno drivers participated considerably in the evacuation of the citizens. In the 1987 assault on Kurmuk the convoy of lorries carrying the evacuated citizens was able to escape due to the good knowledge of the two Abbakars (Abbakar Kharrim and Abbakar Muhammad Bello) of the paths they had once opened through the bush.[49] During other intermittent assaults over the following ten years (1983–97), many of the Maiurno drivers suffered loss of life and property. Now their businesses in that area have been reduced considerably, as a large part of the region is still, at time of writing, in the hands of the rebels.

The Decline of Lorry Driving in Maiurno

Many factors have contributed to the decline of lorry driving in Maiurno and the eventual loss of the drivers' glorious image. Some of these factors are external in the sense that they are not directly related to lorry driving, although they affected it considerably. Other factors are directly related to lorry driving and the driver, and are therefore classified in this study as internal.

---

49. Detailed descriptions of the escape were provided by Abbakar Suleiman (Kharrim), sixty-nine years old, born in Maiurno, interviewed in Damazin on 26 February 2002 and Abbakar M. Bello, sixty-nine years old, born in Maiurno, interviewed in Damazin on 27 February 2002.

The external factors

The main external factors behind the decline of lorry driving relate to the general economic deterioration that Sudan has been steadily undergoing for the last twenty years. This deterioration has been caused by both natural and man-made disasters, in addition to internal and external unfavourable political changes. This is, of course, without mentioning the general economic setback that the whole world has witnessed, be it in the Western countries or in the so-called Third World countries.

For the last twenty years Sudan – along with many other countries of the Sahel zone – has been struck by waves of drought and desertification. These disasters have affected its fragile economy, which is predominantly based on rain-fed agriculture. Any decrease in sorghum, sesame and gum arabic production leads to an equal decrease in the sources of work for lorries, i.e. transportation. The situation has been aggravated by the international fall in the prices of the three major cash crops of the country; namely, cotton, gum arabic and sesame.

Political misfortune is also one of the main factors behind the economic stagnation in Sudan. Shortly after the military junta of Jaafar Nimeiri came to power, the administrative machinery of the country started to lose its vigour, as it opened the door for the general mismanagement of the national finances. By the mid-1970s, Sudanese citizens had already started to suffer from hardship (devaluation of the Sudanese pound, price rises, a scarcity of necessary supplies and deterioration in educational and health services). This period coincided with the boom of '*ightiraab*' in the Gulf countries, where skilled Sudanese were in high demand to provide labour and administrative expertise. The majority of trained and competent civil servants left the country to be administered by those who were less qualified and who could not get jobs abroad.

Although the civil war started as early as 1955, it worsened at the beginning of the 1980s reaching its zenith in the late 1990s and early 2000s. It has exhausted the major part of the already meagre national resources. With the exception of a few privileged sectors of the society, every Sudanese individual has been dramatically affected by the war and it has been disastrous for the Maiurno lorry owners and drivers. As one informant stated, the occupation of the Southern Blue Nile (the main region of their activities) by the rebels 'has brought a complete end to them (i.e. the Maiurno lorry drivers)' ('*intahat minnahum tamaaman*').[50]

The unfortunate choice of the Sudanese government to support Iraq in the 1991 Gulf War made Sudan lose the sympathy of the rich Gulf countries and their hitherto continuous and unfailing financial aid. This, coupled with the accusation that the country is hosting terrorism and violating human rights, has exposed it to all kinds of economic pressures by the Western countries, especially the USA (instigation of the IMF against it, cuts in development aid, unavailability of loans and economic embargoes).

The general economic setback engendered by these factors has affected all aspects of life in Sudan, including the transportation business, on which the value of the

---

50. Ahmad Omar Muhammad (see note 45).

driver primarily depends. However, it should not be assumed that there have been no positive changes in Sudan during this period. In fact, there are some areas that have witnessed remarkable development and progress. More importantly, for the purpose of this study, the areas related to education, the building of asphalt roads and the introduction of modern means of transport have seen significant changes. It is paradoxical that development and progress in these specific areas have also affected the lorry-driving institution negatively. The value of the lorry driver depends essentially on the remoteness of the society with which he interacts from places and societies that might be considered more modern and cosmopolitan. The less modern and cosmopolitan the society, the more valuable he becomes, and vice versa. This is evident from the unfriendly relations claimed by all our driver informants to have existed between them and the passengers on the route between Medani and Khartoum. In the 1970s many of the Maiurno drivers avoided carrying passengers in this region. Unlike the Southern Blue Nile region, where the driver moves and stops at his convenience, in the Medani-Khartoum region passengers are conscious of time. It is here that the driver hears the question that he hates most, that is, 'At what time do we arrive?' Such a question, for him, is a kind of interference in destiny, 'for it is only God who knows if they will arrive or not'. Passengers in this region 'have no respect for the driver: they treat him like a normal person', complained a driver informant. For the same reasons, students constitute another group of unpopular passengers.

In the 1970s, the unfriendly atmosphere was confined to the 'civilized' urbanized area of the Gezira and further north. However, 'civilization', since then, has been progressively pushing its way towards the peripheries, including Maiurno and beyond it, through widespread education, more efficient means of communication and other aspects of modernity (for example, television). Unlike the findings of Duffield (1981: 109–10) in 1977, nowadays one hardly ever sees children pretending to drive lorries or hears songs eulogizing lorry drivers.

Another external factor that has contributed to the drivers' loss of their former status in Maiurno was the appearance of two new competing social classes: the educated people (university lecturers, government employees, university students) and the '*mughtaribiin*' (those working in the Gulf oil countries). While the former surpassed them in outer appearance, the latter surpassed them also in affluence.

Moreover, most of our informants agreed that the recent expansion of asphalt roads and the introduction of comfortable buses and minibuses for passengers and heavy trucks for carrying loads also played a great role in decreasing the value attributed to lorry drivers.

The internal factors

There are many internal or direct factors that are responsible for the reduction of prestige for lorry drivers and subsequent 'devaluation' of the lorry driver in Maiurno. The most important factor emphasized by all our informants is the tremendous increase in the number of drivers, which made the 'supply' far greater than the 'demand'. 'Formerly,' remarked an informant, 'if a lorry owner sacked a driver, he might not find a replacement for a long time. Now he finds ten of them waiting.' In addition to their large number, the present-day drivers are said to be of a lower quality compared with the early generations of drivers. They do not receive sufficient training and they obtain driving licences very easily. Former drivers used to undergo a long and tough process of training whereby they started as junior lorry assistants for about five to seven years, after which the successful ones among them were promoted to senior lorry assistants. At this stage their duty would include, besides taking care of whatever was in the lorry (load, passengers, the lorry belongings, etc.), checking the engine oil, warming up the engine in the morning, tightening the loose nuts and bolts after every trip, helping the driver to repair any major faults and, at a late stage, driving the lorry to a water source to wash it. So, by the time this assistant became a full-fledged driver, he had already acquired the basic mechanical skills as well.

Another internal factor underlined by our informants is the inability of today's driver to comply with the ethics and values of lorry driving. His meagre income does not allow for the conventional hospitality and cooperation vis-à-vis his fellow drivers. Spare parts have become too expensive to be released freely for a broken lorry, and jobless drivers have also become too common to be financially assisted as before.

Conclusion

In conclusion, we have seen that the formative period of the Maiurno society witnessed a series of turbulent events, and life in that period was not as simple as our informants believed. The early years were characterized by flux and change, in which the lorry and the lorry driver acted as both channels and symbols of modernity. As a distinguished class, the lorry drivers maintained the brightest image in the eyes of Maiurno society for a period of about twenty-five years (c. 1960–85). The decline of lorry driving and the eventual eclipse of the driver's star afterwards resulted from a number of external and internal factors, including the spread of education, economic deterioration in some areas and its progress in others. The coming of the civil war to Southern Blue Nile was a final (and maybe fatal) blow to the Maiurno lorry-driving institution.

Social classes are never constant. They emerge and develop under certain conditions and die out with the change of these conditions. Today, the lorry drivers in Maiurno do not stand out from the members of the other socio-economic sectors. Efficient lorries have become very few, while the running cost of old lorries is too high. Therefore, many drivers have abandoned their occupation. Some of them have taken up simple farming (usually regarded as a last option); others retail second-hand spare parts on tables in Sennar and Damazin; while others remain idle, surviving on (irregular) remittances from their sons. There is no chance to uphold the conventional ethics and values even among the few drivers who are still working. So, briefly speaking, one can conclude that the institution of lorry driving in Maiurno has undergone a near total collapse.

Chapter 7

Mbororo (Fulɓe) Migrations from Sudan into Ethiopia

Dereje Feyissa and Günther Schlee

Introduction

This chapter examines the Fulɓe in Sudan and Ethiopia, and particularly the Mbororo among them, who are fairly recent arrivals from West Africa. It explores the way in which the Mbororo have, since their arrival, had to solve a number of problems. The first is that of how to become integrated into a wider system. The Fulɓe left behind in West Africa a set of multiply interdependent pluri-ethnic societies in which they played a number of historical roles and had their own economic niches. Some of the main political units there had Fulɓe rulers; in other settings, members of the numerous subgroups of the Fulɓe were pastoralists, agro-pastoralists or just herdsmen; in yet other settings they were urban traders or Islamic scholars. Fulɓe had been leaving West Africa over centuries, and settling along the route of the haj in North-Eastern Africa and the Holy Cities in Arabia, but a mass migration was triggered by the British conquest of northern Nigeria in 1903. The British were ubiquitous in those days, and difficult to escape, so that, paradoxically, these refugees (*muhaajirun*, participants of a hegira) from the British found themselves, on their arrival in Sudan, in a country that had also recently been conquered by the British. Pastoral Fulɓe migrants, who came to be collectively known as Mbororo, continued to flow into Sudan until the mid-twentieth century. On their arrival, all these migrants found themselves in multi-ethnic settings that differed from the ones they had left behind. They had to carve out new niches for themselves, and had to find new roles in a setting in which professional specialization is shaped partly by ethnicity. In some cases, individuals were able to use the knowledge they had come with: Islamic scholars remained Islamic scholars. Others took up entirely new specializations, such as that described by Abu-Manga in this volume: the Fulɓe lorry drivers. But many of the Mbororo are agro-pastoralists, who used to herd their cattle over relatively narrow ranges of transhumance in West Africa. In Sudan, they seem to have specialized in an 'extreme' form of highly mobile cattle nomadism: they keep moving quickly and ahead of the Arabs to fresh pastures, and move to pastures in the south beyond those frequented by the Arabs. These mobile Mbororo are the subject of the present contribution.

The second and related problem is that of identity. The ways in which their identifications have been produced, maintained and perceived are interwoven with the ways in which they have been integrated into the Sudanese pluri-ethnic political economy. As soon as the migrants found a slot that they could occupy in the new economic and social environment, they had to claim it, assert it and maintain it as a group. This involved making plausible claims to special abilities, avoiding competition by doing something different from others and letting that specialization become so much part of themselves that others would not contest it. They had to shape their customs and habits in such a way as to be similar enough to the host society or the more powerful among their neighbours not to be exposed to too much hostility, and they had to be different enough to be allowed to have a special economic niche.

The third problem can be named after the title of these volumes: changing identifications and alliances. The migrants quickly became caught up in existing power games and the conflicts that surrounded them. During the civil war in Sudan, for example, some Mbororo maintained a peaceful and evasive strategy. They interfered as little as possible with others and moved on fast wherever there were problems. The war situation encouraged others to become 'warrior-herdsmen', like so many other pastoralists. As West Africans in a country that was becoming increasingly polarized between 'Arabs' and 'Africans' and between 'northerners' and 'southerners' (all of whom often looked alike), the Mbororo have, since the 1990s, become closely associated with the north and have thus had a precarious existence in the southern reaches of their range.

As the circumstances changed, their weapons changed also. The Fulɓe at one time and in one situation made use of the fear that others have of their magical abilities and instrumentalized the stereotypes that others hold of them. In other situations, they acquired guns and organized themselves into militias like everyone else around them.

The expansion of pastoral Fulɓe, often described as stretching across West Africa, stretches in fact across the entire width of the Sudan belt, well into Ethiopia (see contributions to Diallo and Schlee 2000). In his research on Gambela, which focuses on Nuer/Anywaa relations (see his other contribution to the present collection), Dereje Feyissa came across pictures of and oral reports about Fulɓe, who came to this part of the western lowlands of Ethiopia last in 1997/98. In 1996, Schlee carried out interviews with nomadic Fulɓe (Mbororo), who had their northern turning point in his research area around Abu Na'ama, Sennar State, Sudan, and he collected information on their seasonal migrations. Those interviewed mentioned migrations across the Rivers Baro, Gilo and Akobo, in other words, well into Gambela, Ethiopia and beyond. They described these migrations as one of their options if inter-ethnic violence in southern Sudan precluded more westerly migration within the Sudanese borders. The seasonal migrations they described had a north-south extension of about 700 km (see Map 7.1).

Schlee met some Fulɓe in Sudan who were able to converse with him in Oromo, a language widespread as a lingua franca beyond Wollega, the western part of Oromia, in the neighbouring areas of Beni Shangul and Gambela, Ethiopia. (Other interviews were conducted in Arabic and later, when Abu-Manga came along up the Blue Nile, in Fulfulde.) The information on the Fulɓe in the Blue Nile region of Sudan and those around Gambela suggests that the Fulɓe in both areas were representatives of the same groups. The question that arose therefore was: why have these cross-boundary migrations ceased since 1997/98? The factors that contributed to this

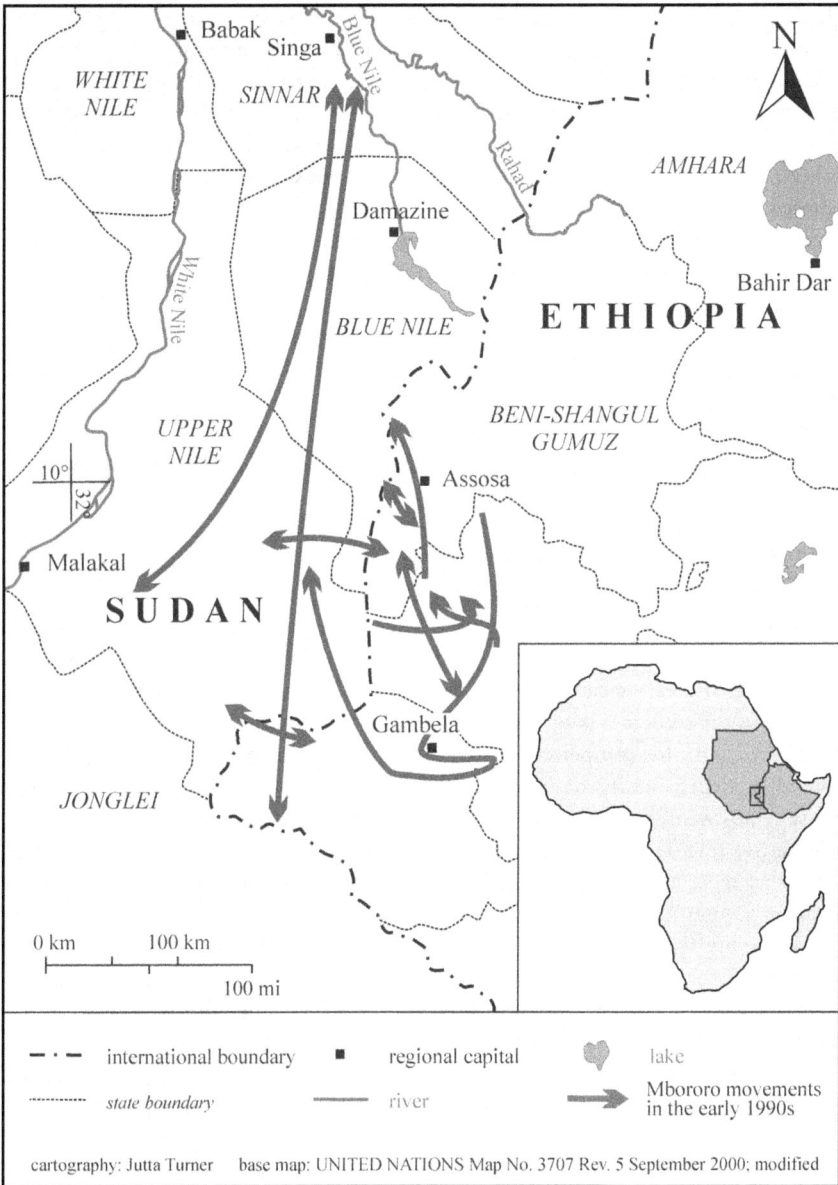

Map 7.1 Mbororo movements in the Sudanese-Ethiopian borderlands in the early 1990s. The two-headed arrows refer to back-and-forth movements, not necessarily precisely along these lines, but in the general area between the two extreme points.

change in migration included: first, the 1997 invasion of northern Sudanese opposition forces from Ethiopian soil into the Kurmuk area of Sudan, and the declining security situation for Mbororo migrants at the hands of the SPLA allies of those northern opposition forces; secondly, a fear of Islamism in Addis Ababa after the attempted murder of Egyptian President Mubarak there in 1995 – the Mbororo and other Sudanese were (wrongly) equated with extremist Islamist tendencies: thirdly, competition for pasture and other resources (possibly market outlets) with local (agro-)pastoralists; fourthly, a strong ecological lobby in the regional governments; and, fifthly, a fear of occult powers and magical qualities attributed to the Mbororo.

For both of us, the Mbororo became a research interest while other research projects were under way: Schlee came across the Mbororo during his research project along the Blue Nile (see Schlee, this volume), while Dereje collected information about their expulsion from Gambela, a theme that related to his research into the politics of exclusion and inter-group relations in Gambela region (see Dereje, this volume). In November 2001, Schlee also collected memories of Fulɓe migrations in western Oromia (Wollega) and Beni Shangul.

The Mbororo in Sudan: Blue Nile Area

Sedentary, urban and village-based Fulfulde and Hausa speakers form a significant proportion of the population of the Blue Nile area. Pastoral Mbororo are fewer, only seasonally present and not so easy to reach because they tend to stay as far away from the cultivated areas as can be combined with the occasional sale of milk by their women in the settlements of the agriculturalists. Their withdrawn life is chosen to avoid the claims for compensation that might otherwise be made following damage caused by their cows in the fields. In this they are not unlike Fulɓe in other areas: those who were not associated with the military aristocracy adopted a very unassertive role (Guichard 1996, 1998, 2000; Boesen 1999a, b). It is only in recent years that these Mbororo and other pastoral Fulɓe of eastern Sudan have acquired a bellicose reputation in their dealings with others.

There are wider and narrower uses of the term Mbororo, and in Sudan and Ethiopia the term is used differently from the way it is used in other places. For example, Guichard found the term used in northern Cameroon to refer to mobile Fulɓe cattle pastoralists in general (Diallo, Guichard, Schlee 2000: 234–36; Schlee and Guichard 2007: 22–23), and Pelican, working on the Cameroon grasslands, found the term used in a similar way to refer to two different sections of pastoral Fulɓe: the Jafun and the Aku (Pelican 2004). Jafun (Jaafun), at least, can also be found in Sudan, but here they do not regard themselves as Mbororo. From the perspective of the pastoral Fulɓe themselves, the Mbororo in Sudan are seen as one section of the pastoral Fulɓe, the Jafun another. In response to the question as to which Fulɓe groups were around Wad an Nail on the Blue Nile at a given time, the following enumeration was obtained: Mbororo, Weyla, Danneeji (*baggar abyad* – 'white cattle'), Uuda, Fallata Malle, Dagara, Booɗi.[1] This does not prevent outsiders from applying

---

1. Abdullahi, the *daamin*, 23 September 2002.

the term Mbororo more widely to refer to pastoral Fulɓe more generally. The oral historical evidence from Ethiopia mostly lacks information about sectional affiliation, and pastoral Fulɓe were there referred to generally as Mbororo or Fallata.

In the literature, Mbororo are described as less Islamized than other Fulɓe (Dupire 1962; 1981; Braukämper 1971, 1992; VerEecke 1988). This may be true at other times and in other places. In Sudan, Schlee witnessed even young people and women doing the *salaat* prayers, and he found no indication of a lax attitude towards Islam. It is, however, claimed that the Mbororo are less literate and less well versed in Islamic practice than other Fulɓe cattle nomads.[2] A Mbororo informant, however, claimed that in the vicinity of his hamlet there was a *khalwa* with a Koran teacher, himself a Mbororo. The adults among this group could not read or write. He himself knew some *suras* by heart. The children did not know much about reading and writing either, but they have made a start. The other tribes around (Jaafun, Booɗi, Duga) are said to be roughly on the same level. Some individuals are literate, others not. Some practices that are rejected by purist urban Muslims are widespread among pastoral Fulɓe. Some of them carry whole bundles of *hiyaab* around with them. Some of these leather pouches, with Koranic writings sewn into them – the Somali would call them *hersi* – are pretty big, almost the size of an A6 pocket calendar. Also the whispering of a spell to a rope, which was meant to prevent by magical means a cow from miscarrying, was observed.[3]

This discourse about different levels of Islamic erudition of different sub-ethnicities of the Fulɓe, is reminiscent of Guichard's (oral communication 1998) and Pelican's (2004) findings about Cameroon, where the Jafun are said to be stricter in Islamic practices than the Aku and the urban Fulɓe. The Huya are said to be more orthodox than both.[4]

The pastoral Fulɓe visited in Sudan include one group of Danneeji ('white cattle') and a cluster of Weyla hamlets. Both groups originally come from Mali and can therefore be included in the wider category Fallata Malle, although by no means all Fallata Malle are Weyla or Danneeji. The Danneeji visited had come to Sudan in the late 1950s. Before that, their last place of residence was Nigeria.[5]

The Migration to Sudan

West Africans, including Fulfulde speakers, came to Sudan over many centuries (Braukämper 1992; Schlee 2000b). The nomadic pastoralists are among the most recent arrivals. Shaikh Baabo ᶜUmar[6] states that his father left the Sokoto area of

---

2. For example, in interview, Abdullahi Moh. Saʾid al Qarawi (14 March 1998) contrasted Mbororo with Jaafun, Booɗi and Duga.
3. West of Wad an Nail, 25 September 2002.
4. See also van Santen (2000: 140ff.) on northern Cameroon.
5. West of Wad an Nail, 26 September 2002.
6. The full text of this interview in Arabic and English, with a transcription in Latin characters to account for the particularities of the Sudanese dialect, has been attached to Schlee's report to the DFG (Deutsche Forschungsgemeinschaft), which can be found under www.eth.mpg.de/Research/Schlee.

northern Nigeria on the arrival of the British. His group moved via Maiduguri and Yarwa into Cameroon, and then into Chad north of Njameena, close to the Lake. They stayed for nine years near the Rivers Saadi (Chari?) and Kuuri (?) and then crossed with Sultan Maiwurno, a descendant of ʿUthmaan dan Fodio, into Sudan, by way of Nyala (see Abu-Manga, this volume). They moved at the pace of livestock, with many breaks, and the whole relocation from Sokoto to Nilotic Sudan took decades.[7] As far as the start of this movement is concerned, these Mbororo can be regarded as part of the Fulani hegira (Schlee 2000b) in the early twentieth century.

Other Mbororo came much later. One man said that he came to Sudan as a small boy in the 1950s from West Africa. He responded to some of Schlee's questions by quoting his father, who once told him that 'the French have written all that down in Niger', suggesting some awareness of Marguerite Dupire's work on the Woɗaaɓe (Dupire 1962).

Outward Identifications: Appearance, Behaviour and Cattle

Contacting the Mbororo was facilitated by the fact that they are easy to identify visually.[8] The lower arms of the women are laden with brass bracelets and their faces are often adorned by blue tattoos. The young men have their hair plaited backwards in thick tresses. On the occasion of marriage the tresses are shorn off. Longitudinal folds remain visible in the scalp for a long time in the places where the skin has been pulled together by tight plaiting.

Mbororo men are also conspicuous from afar because of their shape and colour. While Arabs and Arabized people in Sudan wear wide, white garments, the Mbororo have a preference for close-fitted vests and for the colour blue.[9] There is a special tailor for them in Damazine who applies shoulder pieces, epaulettes, ornamental stripes and decorative buttons to their vests, which give them a military look. Married men have plainer dresses than the youngsters, but they still have a rather close fit. All these are conspicuous deviations from the standard northern Sudanese Muslim dress code.

A factor that did not facilitate the research was a certain level of shyness or evasiveness on the part of the Mbororo. There is no better way to illustrate this than by referring to the field diary in which some of these encounters were recorded. On one occasion, during the rainy season, when the researchers had hired a tractor to get to the locality where Mbororo were said to be, the herds of cattle were driven away just as they approached. An old man under the shade of a tree started ablutions for his ʿasr prayers just as they arrived, apparently trusting in not being molested while praying. After the prescribed prostrations he took his rosary and mumbled 'God help

---

7. Shaikh Baabo ʿUmar, Damazine, 23 July 1996.
8. A full chronological account of the research, closely following the field diary and comprising many data supplementary to those which can be presented here, including maps and pictures, will be published online and on CD by the Max Planck Institute for Social Anthropology, PO Box 110351, 06017 Halle/Saale, Germany, www.eth.mpg.de.
9. There is a beautiful example of the silhouette as a cultural marker in Klumpp and Kratz (1993: 204: the Maasai female silhouette).

me' in Fulfulde over and over again. Fortunately, the presence of Shaikh Baabo ᶜUmar, their spokesman in Damazine, convinced them that we were no great threat, and more and more people gathered around us.[10]

On another occasion, ᶜAwad Karim, a helpful younger colleague from the University of Sennar at Abu Na'ama, and Schlee tried to contact some Mbororo in Wad an Nail. Mbororo women selling milk from house to house were easily recognizable from their naked shoulders, bead ornaments, brass bracelets, tresses and the milk containers on their heads, but they had little Arabic and were very shy. Their evasion of the researchers and a 'helpful' mediator who was encountered in the street (this happened in a southern Sudanese neighbourhood) provoked condescending laughter from the local women.[11] This evasiveness or withdrawal is a significant strategy employed by the Mbororo, and is revisited below, in connection with the relationship between pastoral Fulɓe, Arabs and riverine farmers.

Social identities in this region are also connected to the different breeds of cattle preferred by a group. The cattle of the local Arabs tend to be a whitish milk-type zebu of moderate size, called Kenaana after one group of their owners. The cattle of their Oromo neighbours in Ethiopia tend to be the small East African zebu. In comparison, the cattle preferred by the nomadic Fulɓe are of a demanding, high-performance breed. They are also zebus (*Bos indicus*), but they are much larger and have impressive long horns.

The Mbororo prefer the cattle breed Kuuri Hamra ('red Kuuri'). Their owners sell them only for slaughter. Arabs or Dinka do not want to acquire them for breeding because they find them difficult to manage. They require a lot of pasture and have to follow the rains. They are not suitable for being kept around villages. The saying goes that 'their bull is close in height to a camel'.[12] Their oxen, used for riding and as beasts of burden, carry substantial loads at a fair speed because of the sheer length of their stride. The white cattle of the Danneeji, also called *danneeji* (the people having been named after their animals), are only slightly lighter in build. The size of Fulɓe cattle and their high demand for grazing have played a significant role in the way the Fulɓe presence in Ethiopia in the 1990s was perceived; we return to this issue in that context.

Relationship to Arabs

In his study of nomadic Fulɓe in Sudan Delmet (2000: 199) describes the disregard the Mbororo show towards the grazing and water management of the nomadic Rufaᶜa al Hoi Arabs, and there is certainly resource competition between pastoral Fulɓe and Arab farmers about access to the Nile. It might be expected therefore that the Fulɓe /Arab relationship would be characterized by hostility and negativity, but the Arabs tend to view the Fulɓe nomads with respect or even silent admiration.

For example, Schlee's colleague, ᶜAwad Karim, stressed repeatedly how the Mbororo treat each other respectfully, and he emphasized that this was in strong

10. 22 July 1996.
11. 20 August 1996.
12. Shaikh Baabo ᶜUmar, Damazine, 23 July 1996.

contrast to the conduct of Arabs. He explained that, if an older man approached a group of people, all Mbororo would get up from their *angarebs*, crouch down and mumble benedictions. Tea is also poured in order of seniority of the recipients. Reportedly, the Mbororo explain this with reference to the fear juniors have of the curse of their seniors. To the Arabs who were encountered during this study, the conduct of the Mbororo seemed to conform to some ideal held but not reached by their own communities, and was regarded with a degree of admiration.

On the other hand, there can also be friction and conflict. On one occasion, the researchers were warned of Arabs on their route, who have been known to rob and beat passers-by. These may be particularly Rufaᶜa al Hoi in the area, who have become impoverished in recent years.

Until recently the Mbororo had the reputation of being peaceful and of avoiding conflicts by all possible means. In their relationship with their northern Sudanese neighbours they followed a strategy of moving on or moving away from potential conflict. When moving away has not been possible, problems have sometimes resulted. For example, during the rainy season (*khariif*) of 1999, one Mbororo hamlet stayed quite close to the main road near Abu Na'ama. They could not move their animals without damaging agricultural fields that had expanded in the area. Unfortunately, one consequence was that a donkey belonging to a woman of that hamlet was hit by a lorry. Movement has been constrained in Sennar state because there has been no proper range management allotting migration corridors, grazing areas and watering places with viable access routes to nomads. Pockets of agriculture are everywhere and access to the Nile is blocked by settlements and gardens.[13] Here, nomads are forced to rely on natural and artificial ponds, which collect rainwater, and these are soon exhausted.

To the south of Sennar, in the neighbouring Blue Nile State with the capital Damazine, the status of the nomads seems better. But, even in the Damazine area, one informant claimed that the Hausa have made gardens all along the river so that there is no access to water for cattle. Further south still, in the *Saᶜid*, it is the same. In other places the Mbororo have to pay Arabs for water, but even this is not reliable. They say that Arabs may sell them water once, but, if they come a second time, they might refuse to sell it to them.

In a social universe in which ethnicity is an organizing principle, there are always people who take up mediating roles and become the spokespeople of a group or people who act as experts about one group for another group. One such person is Abdullahi, a Rufaᶜa Arab that Schlee and ᶜAwad Karim met in the livestock market of Wad an Nail. He is the *ḍaamin* for the pastoral Fulɓe, or the 'guarantee man', as ᶜAwad translates it.[14] A security officer who had questioned the researchers at a bus stop had already pointed him out as the Fulɓe *sheekum*, or 'their shaikh'. Abdullahi explained that he had had dealings with the Mbororo since 1963. Every *khariif* season they come to Wad an Nail and, as *ḍaamin*, he guarantees that the animals they

13. ᶜAwad Karim.
14. See contributions to Schlee (2004b).

sell are not stolen. He claimed to speak Fulfulde perfectly, and he showed a significant level of knowledge of their sections and migratory routes.

Relationship to Southerners

The Mbororo hamlets encountered west of Wad an Nail in 1996 did not, in the preceding years, have fixed seasonal routes. The Mbororo decided whether they should move into Upper Nile Province or into Ethiopian territory according to the conditions and safety of the pasture. In southern Sudan, they were involved in frequent conflicts with the Dinka, who captured some of their cattle. The Dinka withdrew fast and far, so that it was difficult to hold them responsible. In one case the Dinka also captured women. The Mbororo then fled from them to Ethiopia. Some of the women later managed to escape and to rejoin their families there.

According to informants, the Mbororo occasionally had to make payments for water to 'Burun, Ingessana, Dawwala, Uduk, Koma and the southerners'. There were also occasions when the Burun and Uduk were said to have shot Mbororo on sight. This made the latter withdraw into Ethiopia to evade Burun and Uduk aggression. Even in the relatively peaceful areas of Sudan, the proliferation of low-level administrative authorities has also made life more difficult for Mbororo. All types of taxes are extorted from them wherever they go.

In the years following 1997, further difficulties shaped the movement patterns of the Mbororo. According to Abdu Omar, the elder brother of Shaikh Baabo ᶜUmar (= Omar), events restricted his group to the area north of Damazine. This resulted from an invasion of northern Sudanese oppositional forces, in alliance with the SPLA in the Kurmuk region, upriver from the research area. Some Mbororo were affected by the fighting, or were cut off on the Ethiopian side and had to hurry to get back into Sudan by going north of the front line. Since this event the route via Kurmuk has not been passable for nomadic Ful6e. Landmines have also been planted there and killings have occurred.

An account collected in Qashmando, Ethiopia, from an impoverished Weyla man who had become a hired herdsman refers to the same period: 1997. He described an SPLA attack against the Fallata in the month of Dahiyya (= Arrafa – the month of the haj), in the rainy season. The SPLA had large camps in the Assosa region, Beni Shangul State, Ethiopia, and the attack also took place on Ethiopian territory. A large number of the hired herdsman's cattle was taken. Other cattle there had belonged to different Fallata sections: Mbororo, Jafuun, Gamba, Weyla, Fallata Malle. All the loading oxen were killed or taken and the women subsequently had to carry the household items themselves. One attack resulted in the death of forty people on the side of the Fallata; another resulted in twelve dead, a figure that included men, women and children. The Ethiopian police did not respond to this emergency; they themselves had tried before to expel the Fallata, and to the Fallata it appeared that the Ethiopian police welcomed the SPLA action. In the following days, the Fallata concluded an agreement with the SPLA. The SPLA were paid 100 sheep in return for the promise to let the Fallata move away from the area without further molestation. The SPLA took the sheep but attacked the Fallata again nevertheless, taking many lives.

In recent years the pastoral Fulɓe have avoided the Kurmuk area and the borderlands around Beni Shangul, by taking more westerly routes. Some recent reports have suggested that, in the dry season (*seef*) of 2002, they have been in ad-Dariyel, an oil area in Upper Nile. The oil people have drilled water wells there and have allowed the herds to drink. Thus the newly flowing oil has affected not only the politico-military constellations but also pastoral movements.

Violent clashes have been reported with Nuer and Burun, and again the Mbororo depict themselves as victims of this conflict. Shaikh Mohammed Saaleh Mohammed AbBakr, the leader of one of the groups that spent the *seef* at ad-Dariyel, complained that the Nuer failed to conform to any law. The Mbororo would go to the *qaadi* court, but he said that, whatever the verdict, the Nuer would continue to take cattle by violent means.

Another elder, Adam Jibriil, explained that fees for grazing and water were payable. His group had been to Girinti, Naasir and Lungushu in the *seef*. They had been using these areas for several years, and there were many Nuer there. Naasir itself was under government of Sudan (GoS) control, but a location called Keek, north of it, was under the control of Riek Machar. Adam Jibriil's group paid fees to the authorities under the GoS or to the Nuer. He described how, every time a Mbororo group entered a new location, new payments were due, normally 500 dinar or one big sheep or goat per camp.

There have been further cases in which the Nuer have rustled Mbororo cattle and killed Mbororo. One camp recently lost one woman and six men. The authorities to whom they paid taxes and fees failed to do anything. Other informants confirmed this picture: payments were made to local powers irrespective of their political affiliation, and they were not rewarded – as one might think taxes should be – by the provision of security or any other government services.

In one hamlet encountered in Khor Dunya, north of Damazine, in September 2002, six people had been killed near Naasir earlier in the year. Some of the Nuer raiders wore uniforms. They looted the camp at ten o'clock in the morning. The youths were not near the settlement at this time, but, when they heard of the attack, they followed the raiders and retrieved some of the stock. In the end, the net loss was about fifty cows and one hundred head of small stock. The day before the raid, the same Nuer had come and asked for small stock for slaughter. The Mbororo had given them nine.[15]

Fulɓe relations with the Burun are almost as bad as Fulɓe relations with the Nuer. In 2002 at Subat (in another version: Daanaja) on the Yabus, the Burun attacked Fallata. The informant himself had a narrow escape. His son was shot through the thigh and taken by a military plane to Khartoum for treatment, where he later recovered and then returned to the camp. Forty-five of the informant's cows were killed. His paternal cousin, who was a member of the *difaʿ ash shaʿabiyy* ('the popular defence'), was killed and his gun was taken. Again, these events took place in spite of the presence of GoS forces in the area.

---

15. Adam Bello, September 2002.

The Gambela Experience

At the time of Dereje Feyissa's research, there were no Mbororo living in the Gambela region. The nearest Mbororo settlement was in Malwal, Mayut Province, on the Sudanese side of the border. The account here is based on the recollections of Gambela residents of the Mbororo stay in that region.

In the early 1990s, a group of nomadic people migrated en masse into Gambela region, the westernmost region of Ethiopia, from southern Sudan. In Gambela they were described as 'Fallata', a generic reference to people of West African origin in Sudan. In this description of what happened in Gambela, the term Fallata is used, following local usage, to describe these pastoral Fulɓe.

In contrast to their previous short visits, in the 1990s the Fallata became highly visible, and even captured media attention. They were presented on national TV as exotic nomads, wandering around in search of pasture. By 1997, however, their image had changed from being seen as a benign and exotic group to one that represented a serious security threat. The change was related to interstate problems between Ethiopia and Sudan. As a consequence, Fallata were harassed by the police and expelled from Gambela by force. Brief as their stay was, the Fallata left a lasting imprint on the social landscape of Gambela. Also, changing attitudes to the Fallata meant that many of their traditional migratory routes and grazing lands were blocked to them. As a result, the Fallata changed their strategies and emerged as one of the key players in the civil war in Sudan. They actively participated in the complex and changing politico-military alliances along the border. As will be shown, whether in the form of intermittent migration, episodic sojourn or attaching themselves to the regional power holders, the 'Fallata factor' is driven by a desire to access grazing land. In the following, some experiences of the Fallata in and around Gambela are outlined in more detail. Types of migration and reasons for migration are explored, followed by a study of the reasons for their expulsion from Gambela town. Prominent representations of the Fallata by Gambela residents are discussed as playing a key role in this expulsion, but the Fallata experience must be understood in the context of the growing politicization of processes of identification and social fluidity in this border area.

Migration patterns, phase 1: pre-1994: intermittent and ephemeral

Before 1994, the migration of the Fallata was restricted to Nuer areas up to Itang. Nuer call the Fallata Paluath, probably a corrupted version of the term Fallata, itself possibly derived from 'Fulɓe'. There were no recollections of major conflicts between the Ethiopian Nuer and the Fallata. They are mentioned, however, in one of the popular war songs of the western Nuer, which recounts their major fights with various groups of people:

> *Kor tilian me dan* (Don't forget the fight with the Italians)
> *Ke kor dec doar* (the fight between Anyanya I and the Arabs)
> *Ke jalab ni kor gong ke cagei* (the fight between Nuer and Oromo)
> *Kor Makeri ke lingelith kamda* (the fight between Nuer and British)
> *Ke Paluath ni* (the fight between Nuer and Fallata)

For the Ethiopian Nuer, especially the Gaajak, who live in Jikaw district,[16] the Fallata appeared exotic, with their camels and their use of cattle for transportation.[17] They were said to have stayed for only two or three days at a time in a particular location, too brief a time to arouse local suspicion. Instead, their arrival was often longed for, as they used to exchange their abundant milk for Nuer grain and other items. This was particularly during the peak of the dry season, in the months of March and April, when Nuer pastoral communities form temporary dry-season settlements along the river and beyond. The youths take the cattle away to far places, while the elderly, women and children remain in the settlements. It is understandable that the Fallata were welcomed by the Nuer during such a lean period.

Migration patterns, phase 2: 1995–97: sustained and highly visible

By 1995, the number of Fallata who had settled at various places in Gambela region was estimated at 15,000. This phase of their migration coincided with the escalation of communal violence, triggered by factionalism in the rebel movement in southern Sudan. The Fallata were said to have been a favoured cattle-raiding target because of their conspicuous cattle wealth. Renewed fighting between southern Sudanese rebel faction leaders in the mid-1990s created a greater level of insecurity along their traditional westerly route of migration (near Assosa), giving them an added incentive to move to the Gambela region. Fallata began to move to the west of Gambela town, well beyond Itang, which was hitherto their frontier. This time around, the Fallata went as far east as Abobo and Pinyudo, and settled amidst Anywaa.[18] Apart from their own language, they spoke Arabic and Oromo. They became highly visible, and their sudden appearance in Gambela made them an object of suspicion, especially as the new regional state was busying itself with identity politics. After about two years, they were expelled from the area.

Dereje Feyissa talked to various categories of people in Gambela town and to villagers in Jikaw and Itang area. He explored the nature of their encounter with the Fallata and the reasons given for their expulsion. With various levels of emphasis, they gave four explanations for the expulsion, in which the Fallata are portrayed negatively: (1) as excessive environmental degraders; (2) as non-payers of taxes; (3) as Islamic fundamentalists, or allies of Islamic fundamentalists, and a political threat; and (4) as people with dangerous and disruptive magic powers. Each of these is explored in more detail in turn:

---

16. Dereje Feyissa is more familiar with the Gaajak Nuer. His dissertation project focuses on the Gaajak Nuer and the Openo Anywaa along the River Baro in Gambela, Itang and Jikaw districts.
17. No pastoralist group in Gambela keeps camel. Similarly with pack animals: neither mules nor oxen are used for portage for climatic and cultural reasons.
18. Informants from Gambela town located the routes of Fallata migration from Sudan to Assosa (Beni Shangul region), via the Dambidolo road (Wallega) to Itang, and from Itang to Jikaw and Gambela and Abobo districts.

Ecological: overgrazing and deforestation

Political officials and professionals from the Bureau of Agriculture in the Gambela Regional Government emphasized particularly the ecological reasons why the Fallata were expelled. According to Okello Oman, then President of Gambela Region State, for example, the Fallata were seen as environmental degraders: 'Those Fallata don't care for the environment. They cut the forest to make in-roads as well as let their cattle overgraze. They even put salt on the grass such that the cattle will graze exhaustively and eat everything, including the roots.'[19]

The settlement of the Fallata was also a subject of discussion at the 1994 conference on 'National Resource Development and Potential Conflict' convened by the Gambela Regional State. The conference recommended that the Council take immediate action, as 'their [the Fallata] animal husbandry has put a heavy pressure on the local eco-system' (Bureau of Planning and Economic Development 1994, vol. I: 56). Their fear was aggravated by the perception of a similar threat from settled Uduk refugees from southern Sudan. The entire Uduk community, estimated at more than 10,000, were resetted in 1992 at Bonga refugee camp, around twenty kilometres east of Gambela town (James 2000c). The Uduk were also considered by the local authorities to be 'ecologically unfriendly'. Against the background of the huge influx of refugees in the 1980s and 1990s and the ensuing deterioration of relations between the refugees and the local people, certain xenophobic attitudes towards outsiders developed. For the larger part of the 1990s, the newly constituted regional government was mainly dominated by the Anywaa, who took certain measures towards empowering 'locals' vis-à-vis 'outsiders' (see Dereje Feyissa in the present collection). The seemingly ecological argument, therefore, encompassed more political motivations in the context of an intense politics of exclusion, which was heightened by the territorialization of ethnicity and the ethnic federal politics organized by the Ethiopian state.

Fallata as non-taxpayers

A main criterion for inclusion or exclusion from citizenship rights is the payment of taxes. One Nuer government official reported that the main reason why the Fallata were expelled from Gambela was because they were not paying taxes. It is significant that the informant was a Nuer official because one of the main strategies used by Nuer immigrants from southern Sudan in order to gain entitlements to natural resources is either demonstrating that they have already paid tax or agreeing to pay tax in return for access. The Fallata do not appear to have tried such a strategy, but, as will be seen in the later part of the chapter, they have employed a similar strategy, albeit in a less formal way, by paying rebel groups or warlords in order to have access to much valued dry-season grazing land.

Despite the discourse of Fallata as non-payers of tax, the general public, and even the officials, also acknowledged the economic contributions that the Fallata made to Gambela town. Their economic activities included the sale of milk and butter by

---

19. Interview, April 17 2001, Gambela town.

Fallata women and the sale of cattle for slaughter by Fallata men. Gambela town is said to have been flooded with milk and butter, and there was also bartering as the Fallata started exchanging milk products with flour. In addition, the entire local demand for beef was met by Fallata alone.

Political: perceived threat from Islamic fundamentalism

Despite the recommendation by the 1994 conference on resource management, the Fallata were not expelled from Gambela region until 1996–97. Their expulsion needed a further trigger and this was supplied by the international geopolitics of the second half of the 1990s. At this time the United States was turning its attention increasingly to the apparent resurgence of political Islam in both the Middle East and Northern Africa. In the 1990s, the government of Sudan was seen by the United States as a 'terrorist state' and one of the main sources of dangerous Islamic fundamentalists. Sudan's connection with Islamic fundamentalist forces became a particularly increasing concern after the assassination attempt in 1995 on President Hosni Mubarak at Addis Ababa airport. Both Ethiopia and Egypt claimed at the time that Islamic terrorists 'harboured' by Sudan's Islamic regime were responsible and demanded that they be handed over. This provided an excellent opportunity for the United States to pursue its policy of isolating Sudan. It managed to get UN sanctions placed on Sudan in 1996, and most of the neighbouring countries, particularly Eritrea, Ethiopia and Uganda, severed diplomatic relations with Sudan. At the same time these countries increased their political and military support for the rebel movements, particularly the SPLA. As a further step, Ethiopia closed down the Sudanese consulate in Gambela, as well as Islamic NGOs operating in Ethiopia that were argued to have connections to Sudan.[20] In order to combat the 'growing Islamic threat', the neighbouring countries received substantial financial and military support packages (Gilkes 1999).

The Fallata became entangled in this international diplomatic story. Rumours in Gambela town spread that the Fallata had come to convert the local people to Islam. It was said that they would only socialize with Muslims and that their leader had made frequent visits to the mosque in Addis Ababa. Their Islamic background was exaggerated and they were discussed almost as if they were working for the Sudanese intelligence service. The words of Okello Oman, then President of Gambela State, show, for example, the way the Fallata were seen as having a long history of being employed in the interests of the Sudanese state: 'The Fallata have been very much identified with the Sudanese state. The British first brought them from West Africa to vote for independence against the Egyptians. That is why they have been allowed to move freely.'[21] These words also reflect the growing relevance of the state in defining who people are. This, in turn, depends on which of the state's interests are at stake, a matter often related to the maintenance of its own power. The result here

---

20. The Ethiopian government even went to the extent of dismissing what it called Islamic extremists from some regional states, especially Beni Shangul and Somali regions.
21. Okello Oman, interview, 17 April, Gambela.

was that what was originally practised as a survival strategy (a production regime based on extensive mobility and transhumance) became politicized. The importance of the international dimension of the Fallata's image, which constructed them as a security threat, is particularly clear in relation to the fact that Gambela Regional State made pastoral mobility constitutional in 1995 but the Fallata were not considered a security threat until the assassination attempt on Mubarak.

Fear of their magical power

The popular image of the Fallata as mystical, as 'other' and as possessing magical powers, has also factored into the expulsion of the Fallata. As the 1990s unfolded, the Fallata's magic was seen as increasingly dangerous. The Fallata have long been considered exotic, with their pack animals (both camels and cattle), their use of bows and arrows and their unique system of animal husbandry. One Nuer informant summarized the general 'otherness' of the Fallata as follows:

> The Fallata people are different. They are the only people who keep camels [in the Gambela region] and carry goods on cattle. They have bows and arrows, with snake poison. They are sharpshooters. Their cattle are also as unique as they are. Their cattle listen to Fallata language. The Fallata travel long distances. They wear special shoes, modified plastic shoes [locally called Congo shoes] along their ankles so that they do not get cut by the elephant grass. The Fallata own more than 200 cattle each. It is difficult to domesticate their cattle. Nuer who get Fallata cattle tame them by keeping them for three days without food and drink or beating them until they fall down.

In a similar vein, another Nuer informant said: 'Even if you manage to raid Fallata cattle, you cannot keep them. The way they keep their cattle is different. Unless you take the calves and rear them as Nuer cattle, they will go back to the owner.'

The Fallata are also seen as highly qualified magicians, and, in many parts of Ethiopia, the very name Sudan is identified with magic. The Fallata are said to have magic spells written in Arabic. According to informants, most of the magic performed by Fallata in Gambela related to love charms or protection. People employed Fallata magic in order to keep themselves safe from offenders and insults. The Fallata were also said to use magic against those who wronged them. One story was related about a certain highlander who had sold them a condom, saying that it was a veterinary drug. The highlander died soon afterwards.[22]

The association of Fallata with magic also becomes clear from the accounts of a civil servant in Gambela town, who had many stories of his encounters with the Fallata.[23] He told of how the regional administration had tried to send the Fallata

---

22. Narrated by Okello Akuway, an Anywaa from Gambela town who was close to the Fallata.
23. Fuller version in Schlee (2008: 19 February).

back to Sudan but that the local population had been afraid to implement the policy for fear that the Fallata might harm them with their magic. He also told of an encounter that he had with a Fallata in a bar, after which the Fallata had given him a love charm. The love charm consisted of various pieces of 'medicine', and he was told to rub his body with the mixture after bathing on three subsequent Fridays. He claimed to have followed the Fallata's instructions and boasted that he had enjoyed remarkable success in his sexual exploits as a consequence.

In another account, Amtallaqa, a Gambela resident who was kind enough to provide us with photographs (see Plate 7.1), explained that someone had called a Fallata 'an animal' in Oromo. The offended Fallata took two nails, the length of Amtallaqa's finger, out of his pockets and inhaled them through his nostrils. He then threatened to retrieve them through the penis of his offender. The latter did not delay in apologizing and clarifying that the Fallata was by no means an animal. He was thus saved the procedure. On another day, a Fallata woman came to Amtallaqa's office and he gave her three birr for tea. She asked him whether she could help him in any way. He replied that his wife had continuous pain in her left breast. The woman left and Amtallaqa's wife never had any pain of that sort afterwards.

Amtallaqa's accounts are similar to those given by other informants. People generally talk about a 'magic boom' in Gambela that accompanied the arrival of the Fallata. Later accounts of this 'magic boom' take on a darker hue, however, as the Fallata magic came to be portrayed as more sinister 'sorcery'. Accounts referred more frequently to the ill effects of magic: the love charms were said to have caused family splits and interpersonal quarrels. These miraculous and exoticizing stories of the magical achievements of the Fallata, told by Amtallaqa and others, tell little about

Plate 7.1 Amtallaqa between Mbororo women

what actually happened. But they reflect the attitudes of Gambela residents to the Fallata, and the degree to which they were portrayed as different and 'other'.

There are also instances when the politicians in the Gambela region were said to have harnessed 'Sudanic' magic to attain positions or to mobilize their constituency. Among the Nilotic communities of the borderlands, groups are hierarchically positioned depending on their access to magical power. The Dinka occupy the upper echelons, followed by the Nuer, while the Jallaba (northern Sudanese Arabs) and the Fallata are considered to be the strongest of the 'red' people. The Nuer claim superiority over Anywaa not only because of their own magic, but also because they are said to 'buy' magic from the Dinka or to persuade Dinka magicians to work in their favour. In such inter-ethnic competition, these instances could be described as 'tapping the irrational' for political action, to borrow Chabal's and Daloz's term (1999). In the confrontation between the Anywaa and the Nuer during the political turmoil of the transitional period, the Anywaa-based GPLM soldiers (Gambela People's Liberation Movement, a rebel movement that took over power from Derg officials in Gambela in 1991) are said to have worn bulletproof magic called *kunjur*.[24] Many Anywaa youths were said to be so impressed by the magical qualities of *kunjur* that they joined GPLM en masse. Such manipulation of magical power may have boosted Anywaa morale in a context in which they were seen as relatively weak compared with other groups.[25] The GPLM also had military camps in Sudan, from which it conducted its activities against the Derg regime. Some informants asserted that the *kunjur* and the idea of using occult power for politico-military purposes were learned from the Fallata. The improper use of the *kunjur*, such as sleeping with women or eating meat and drinking alcohol when the magic is still active, was said to have driven many GPLM soldiers mad.

In summary, the politicization of the Fallata presence, coupled with their association with occult powers, generated a general atmosphere dominated by a 'fear of the unknown', which further contributed to their expulsion from Gambela. Whatever the reason(s) for their expulsion, the Fallata were harassed and finally chased by the police from the Gambela region. They were targeted so much that it was said that they became scared of anybody holding a stick, fearing that he could be a policeman. There was no one who championed their cause or defended them. In spite of their popular connection to Nigeria, the Nigerian Embassy in Addis did not recognize the Fallata as its citizens. Discussions with the Nigerian Embassy staff revealed a strong modernist discourse in which the nomadic Fallata were seen as lawless and without state identity. For them, the Fallata were presented either as peaceful and interested in sedentarization and acquiring a modern life, or as warrior-like people who do not pay taxes and 'wander around', creating a security threat in

---

24. Similarly, EPRDF soldiers were believed to have possessed 'Sudanic magic' when they entered Addis Ababa in May 1991 with donkeys and other pack animals. They seem to have encouraged such mystification among the populace as they were not sure of its loyalty.

25. There are numerous Nuer magicians, prophets and sorcerers. The Anywaa particularly acknowledge the power of and fear what they call 'Nuer evil force', called *wel*. They believe that, if they touch or kill Nuer cattle and eat them, *wel* would attack them.

their host countries. From the viewpoint of the Nigerian state, the Fallata, whom they called the Mbororo, who left Nigeria a long time ago through Cameroon, are troublemakers who do not recognize state boundaries.[26]

Migration Routes

Travelling through western Wollega and Beni Shangul in November 2001, Schlee enquired in a number of localities, with the assistance of Getinet Assefa, about the last visits of the Fallata. They met a Pullo (sing. of Fulɓe) who had become a hired herdsman of a local Oromo shaikh. The accounts from this area provide a more differentiated picture of the routes taken by the Mbororo than those obtained from the Gambela informants.[27] The routes described by informants in Sudan, and locations remembered as having been frequented by Fulɓe pastoralists by people in Ethiopia have been synthesized into Map 7.1.

The attitudes towards the Fallata of the local people in western Oromia corresponded closely to those of Gambela. Their herds were described as large and the appetite of their cows as enormous. For example, in Dambi Dolo they were last seen in 1993. Below Dambi Dolo, in the Qeeto area, where they stayed most of the time, the local population complained about pasture degradation. Mbororo cows were said to be very destructive, even eating the high, hard savannah grass, stem and root, which is not consumed by the local cows.[28] Here also there have been many attempts to expel them.

At the same time there was also market exchange. The Mbororo sold a lot of butter. In Dambi Dolo, in spite of the rural character of the town and the fact that many people kept some cows even in town, it was difficult to buy milk. The Mbororo therefore found a ready market there for their milk.[29] In other places, people were reluctant to buy milk from the Mbororo for fear of their magic.[30] Generally, oxen were sold for slaughter or for breeding stock, although there were many reports that local people failed to succeed in breeding Fulɓe cattle. The meat was sold more cheaply than that of local Oromo cattle and had a different taste. The butter was said to be watery and only fetched eight birr/kg as compared with ten to twelve birr.

In Beni Shangul, close to Assosa, pastoral Fulɓe had been seen in the area as recently as 1999, but they were only men and boys. The rest of the families were left on the Sudanese side, in the Yabus area. The men and boys had 3,000–4,000 cattle with them. They came to the village to sell milk, cattle for slaughter and even heifers for breeding. The animals they sold died later in an epidemic, however (the Mbororo cattle are not disease-resistant), but some of their mixed offspring from Berta bulls remain with the Berta.[31]

---

26. Nigerian Embassy, Addis Ababa, May 2001.
27. The original data can be found on the CD, which contains the full documentation (for more information consult www.eth.mpg.de).
28. Dr Ephraim Mamo, 27 November 2001.
29. Ibid.
30. Gashaw about Muggi, Ambissa about Shabal, 20 November 2001.
31. Babikir ʿUmar, in the Berta Village Abraamo/Ibraama, 24 November 2001.

The New Militants

Generations of students of Fulɓe culture have tried to capture the essence of Fulɓeness, of *foulanité*, as some call it, with varying degrees of irony. What is the common denominator of all the different Fulɓe groups, nomadic or sedentary, rustic or urbane, illiterate or erudite, which stretch from Senegambia to Ethiopia across the width of Africa? Native concepts such as *pulaaku* or *semtende* have lent themselves to such discussions, eagerly taken up by Fulɓe intellectuals. These concepts have something to do with self-denial, modesty, toughness, with the virtues of the herdsman in the dry savannah. But there is more than one version of this core value. In settings where the Fulɓe have had to adjust to a political power other than themselves, there has been a stress on keeping emotions in and on withdrawing and avoiding conflicts (the modesty aspect of *pulaaku*). In other settings, as in the successor states of the Fulani jihad, where Fulɓe have or had the upper hand, their central values did not differ much from the ideology of other warrior aristocracies (the toughness aspect of *pulaaku*). The following quote from a Mbororo man suggests that, in the Sudan-Ethiopia borderlands, the Mbororo have experienced a shift from the former to the latter:

> There are SPLA attacks against us every year. In one year there are ten dead, in another five, from Weyla alone. In the south, one can no longer be sure of one's life. At any time one can be shot or deprived of one's cattle. Fallata are known for their ability to suffer without lamenting, but what can one say now?[32]

The words portray an increasing level of suffering and the sense that the Mbororo have felt that a violent response is therefore increasingly legitimate. Overall there has been an increase in levels of violence on all sides. The story is the same as in many parts of the world, and it has been made possible by the huge influx of AK-47s (Kalashnikovs).

In this situation, new forms of organization have also arisen for the Mbororo. A young Mbororo man showed us[33] his membership card in a militia. It was called *Katiibat al haqq al muʿminiin* – 'Militia for the right of the believers'. The card was shrink-wrapped in plastic. This militia consisted entirely of Mbororo. Its leader was called Baabikir Barka and it was based at Agade.[34]

The Weyla have also had their own militia, whose leader was Saaleh Bank. Saaleh Bank was killed in 2002 (probably in June) near Malkan, in the Funj region, by the SPLA. Saaleh Bank had fought more than a hundred battles, was rich in cattle and was supported by the government of Sudan. Saaleh Bank appeared in Mbororo accounts as a legendary figure. For example, in a camp of Weyla, another nomadic Fulɓe group, west of Khor Dunya, a man who had just come back from the fighting reported how Saaleh and his group had run out of water and become encircled. His account

---

32. AbBakr Ahmad Jibriil, West of Khor Dunya, 2 October 2002.
33. Schlee was assisted by El Hadi Ibrahim Osman and ʿAwad Karim.
34. West of Khor Dunya, 28 September 2002.

emphasized that there was no bullet hole in Saaleh's body, but that he was struck down by the explosion of a bomb. This detail is significant locally because it does not conflict with the belief that Saaleh was bulletproof because of strong magic. The magic, of course, was of little help as his death was caused by something other than a bullet.

These pieces of information illustrate the extent to which warlike attitudes had, by 2005, come to replace the Mbororo's earlier strategy of withdrawal and evasion. Further research is continuing in order to fill some of the gaps in this picture.

Conclusion

Deprived of the possibility of peaceful migration, the Fallata became deeply involved in the local system of politico-military alliances and conflicts along the Ethio-Sudanese border, in order to promote their own interests. They have been involved in small-scale arms trafficking and actively involved in the inter-factional fights among the rebels in southern Sudan, organized through a set of political clientelist networks. The pattern of alliance in the politics of liberation in southern Sudan was shaped by the collapse of the so-called Sudan peace agreement, which lured a significant faction of the former SPLA to the government side. In January 2000 Dr Riek Machar and his Nuer-based SPDF (Southern Sudan People's Democratic Front) withdrew from the peace agreement. By April 2000, his forces resumed military activities from the Mayut area along the border. As a response to such moves the government of Sudan armed a Nuer warlord called Gordon Kong, who established a movement called Thawra Jikany (the so-called Jikany Revolution). Thawra Jikany claims to represent the Jikany Nuer (also called eastern Nuer) as a counter to Riek Machar's predominantly western Nuer (Bentiu) constituency. In February 2001, Mayut district was the battleground between the forces of Riek Machar and Thawra Jikany. As a countermove Riek Machar intensified his movement's link with one section of the Jikany Nuer, the Gaajak, while the Thawra Jikany secured the support of part of the Gaajok section.

The Fallata were caught up in this renewed civil war, which drew more and more local groups into its orbit. In the confrontation between Thawra Jikany and the SPDF, the Fallata allied with the former. The alliance between Thawra Jikany and the Fallata was based on mutual interests. The Fallata needed access to dry-season grazing land, especially after they were excluded from their traditional migration routes in the direction of Assosa (Wollega/Beni Shangul) and Gambela after 1997/98. They also needed patrons to protect them from rebel attacks as well as from communal raiding. The Fallata were the preferred prey for both because of their cattle wealth. As the civil war expanded its horizon, the Fallata realized the need to tap their 'irrational' (occult power) and also recognized the realpolitik and armed themselves for protection. On its part, Thawra Jikany demanded a contribution from Fallata (mainly in terms of cattle) in order to allow them to have access to dry-season grazing land. The Fallata were said to have come to the Malwal area near the Ethiopian border from the Omdurman[35] area via Yom district. In return for their settlement at

---

35. This cannot mean the part of Khartoum known by this name. There is also a place called Omdurman Fallata on the Blue Nile.

Malwal, they paid Thawra Jikany eight cows per day. Malwal was the main dry-season grazing area for the Gaajak Nuer (particularly the Cieng Nyajani and Cieng Wau sections). The Cieng Nyajani did not like the presence of the Fallata in Malwal, all the more because they inhabited the most resource-poor areas except for Malwal. The situation worsened following the beginning of a protracted resource conflict between the Cieng Nyajani and Cieng Wau, which rendered large areas inaccessible. The Cieng Nyajani raided the Fallata in February 2001 and took around 800 cattle and a lot of goats. In order to create a strong sense of dependence on the side of the Fallata, Thawra Jikany created an atmosphere of fear and anxiety for them and continued to represent them as antagonistic to other Nuer and as more vulnerable to cattle raiding.

It is becoming clearer that the relative strength and weakness of groups in the conflict zone is partly determined by how successful they are at attaching themselves to the regional power brokers. Whether the Fallata have been able to successfully manipulate the powerful groups or whether they have been weakened through their involvement in the civil war remains to be seen. But one thing is clear from the unfolding story: while the GoS and the rebels continued to portray the war as one between the 'Arabic north' and the 'African south', there were and are many other wars going on at different levels. For the government and the rebels it was a question of sharing or monopolizing power, while for the local agro-pastoralist groups it was all about access to pasture and cattle raiding. The conflict and alliances can therefore be seen as a case of 'old wine in new bottles': the motivations were the same, but the rules of engagement and alliance had changed. The net result was the politicization of the fluidity of identity along the border, mainly expressed in terms of increased

Plate 7.2 Mbororo herdsman, Blue Nile area, 2002 (photo G. Schlee)

insecurity and restricted mobility. The image of the Fallata as 'peaceful' changed. In the context of clientelist networks, neutrality became a political liability. In their search for patrons the Fallata became caught up in the civil war in Sudan on both sides. The government of Sudan possibly played up their shared Islamic connections, while the Nuer warlords played the Fallata off against their competing neighbours. As the civil war dragged on, there was an increasing shift from 'communal violence', giving the warlords more legitimacy to continue their activities and providing them with the leeway to manipulate inter-group relations. Previously, the Fallata were considered as neither Jallaba (Arabs) of the north nor black of the south, but in a category of their own like other 'red people' of south Sudan. Their increased involvement in the civil war on the side of the government and government-supported rebel groups has meant that they became caught up in an increasingly polarized discourse about identity in the political landscape of Sudan. They have come to be seen as much more firmly situated with the Arab north and against the black south.

Part IV
Displacement, Refuge and Identification

Chapter 8

Conflict and Identity Politics: The Case of Anywaa–Nuer Relations in Gambela, Western Ethiopia

Dereje Feyissa

Introduction

This chapter is an attempt to explain an emerging regional pattern of conflict and identity politics, focusing on their occurrence among Anywaa and Nuer, the two main ethnic groups in Gambela region, western Ethiopia. In this chapter, I explore identity as a political process through which the strategies of exclusion/inclusion are articulated. Such an approach brings in complex actors in the identity game, within and beyond the ethnic unit. Contrary to the 'traditional enmity' approach or the emerging essentialist public discourse about ethnic relations, I argue that the hostility between the Anywaa and the Nuer is a 'modern hate' and that the resultant conflict is driven by the wider political changes in both Ethiopia and Sudan. Particularly influential is the encroachment/expansion of the modern state at the periphery and the changing system of alliances and notions of belonging that accompany it. The new political processes have also activated a pool of collective memory, especially related to the historical eastward pastoral expansion of the Nuer.

Introducing Gambela

Gambela region is located in the westernmost part of Ethiopia along the south Sudanese border. The region is the home to speakers of a variety of Nilo-Saharan languages, such as the Nuer, Anywaa, Majangir, Komo and Opo. Differentiation among these social groups is also evident in their different livelihood strategies, with varied emphasis on cattle-keeping (Nuer), riverain cultivation (Anywaa) and shifting cultivation and hunting/gathering (Majangir, Opo, Komo).

Gambela and the neighbouring south Sudan regions have experienced three great Nilotic migrations (Johnson 1986). The first wave of migration was in the seventeenth century by the Lwo group of the Western Nilotes, which comprise the Anywaa, the Shilluk and the Luo (Collins 1971). The Anywaa,[1] together with the

---

1. The Anywaa are variously spelled as Anuak, Anywa, Anyuak, etc.

Shilluk, took a north-westerly direction probably from the central parts of southern Sudan or the south-eastern region near Lake Turkana in Kenya, and settled along the major tributaries of the Nile in what today are the south-eastern part of southern Sudan and the Gambela region of western Ethiopia (Perner 1994). This was followed by the Dinka migration of the eighteenth century and the massive expansion of the Nuer in the nineteenth century (Collins 1971). Most of these areas were later taken over by the Nuer, who migrated from their homeland west of Bahr el Jebel in southern Sudan, through a combination of conquest and effective assimilation, beginning in the 1820s and continuing to the present (Jal 1987). The Nuer who live in the present-day Gambela region belong to the eastern Jikany branch, consisting of the Gaajak, Gaajok and Gaagwang Nuer.[2] The majority of the Nuer live in the Upper Nile region of southern Sudan; most of the Anywaa live along the four major rivers of Gambela – Baro (Opeeno), Akobo, Gilo and Alwero – and a small minority live in Pochala district in the Upper Nile region of southern Sudan. By 1930, there were already groups of Gaajak Nuer who were permanently settled in Ethiopia, in addition to a more sizeable group who crossed the boundary in search of dry-season grazing grounds (Johnson 1986; Hutchinson 1996).

According to the 1995 census, Gambela region has a population of 182,000 in a land size of 25,274 square kilometres. The Nuer constitute around 40 per cent, while the Anywaa are estimated at 27 per cent of the region's population.[3] At present the Anywaa live in six of the nine districts of the region, while the Nuer predominantly live in the two densely populated border districts of Jikaw and Akobo.

The complexity of the social composition of the region was increased as the expansion of the Ethiopian state into the region at the turn of the nineteenth century resulted in new population movements. First, and over a long period of time, there was the spontaneous migration from the neighbouring Oromo-speaking highlands; and, in the early 1980s, the government resettled a large number of drought-affected farmers from northern and southern Ethiopia. Today these people are referred to variously as *gaala/degenya* by the Anywaa and *buny* by the Nuer, a social-ecological category denoting phenotypic and cultural difference between the lowland Nilotic people and immigrants from the highlands (hereafter referred to as highlanders).[4] According to the census, the highlanders constitute around 24 per cent of the region's population.

Various criteria used for distinguishing between these communities can be discerned. The most visible line of differentiation is between the 'indigenous' Nilotic communities, on the one hand, and the highlanders, who speak various Semitic and Cushitic languages, on the other. The criteria of distinction are knowledge of and

---

2. These Nuer groups can also be written Gaat-Jak, Gaat-Jok, and Gaat-Guang. The above spellings are used for consistency with other chapters.
3. Both the Nuer and the Anywaa bitterly dispute the results of the census, which allegedly underestimates their respective population size. Both claim that areas not covered by the census would make a significant difference to the demographic balance. In addition, the Anywaa consider the population size of the Nuer to be inflated by a recent influx from southern Sudan.
4. Gambela region is 500 masl, sharply contrasted with the neighbouring highlands, which rise well above 2,000 masl.

competence in the national culture and skin colour: the 'dark-skinned' Nilotes are contrasted with the 'light-skinned' highlanders. This perceived physical boundary is also associated with power relations: the highlanders dominated the regional politics up until the establishment of the Gambela regional state with the new ethnic federalism in Ethiopia in 1991. Post-1991 Gambela has seen a reversal in power relations in favour of the local people, so that now 'indigenous' people fill political positions and senior administrative posts. The highlanders are politically marginalized but they still remain 'significant others' in inter-ethnic relations and the political debate between the Anywaa and the Nuer that has ensued.

Conflict and Identity Politics

The Anywaa and the Nuer are the key players in the current identity game in the Gambela region. The conflicts that have arisen between them have taken different forms: conflict between the predominantly Nuer refugees from southern Sudan and the local Anywaa; armed conflict between Anywaa- and Nuer-based organizations; rivalry between Anywaa and Nuer political parties; the split of a local Protestant Church into Anywaa and Nuer congregations; violent clashes among Anywaa and Nuer students; and communal violence among villagers. Nowhere else is the tension reflected more than in the regional town, where the ongoing 'ethnification' process has been physically manifested in exclusive residence patterns.[5]

Earlier studies on Anywaa-Nuer relations characterized conflict as embedded in the 'social structures' (Evans-Pritchard 1940b; Sahlins 1961; Kelly 1985) or lack of interdependence (Shumet 1986), or emphasized the external dimension: the emergence of Nuer-based rebel groups in the war in southern Sudan and the resultant cross-border ethnicity (Kurimoto 1997, this volume). The Anywaa-Nuer relationship is described as 'ancestral enmity' (Bahru 1976; Perner 1994) or in conspiratorial terms (Yacob et al. 2000). In a similar vein, the public discourse in contemporary Gambela represents the conflict between the Anywaa and the Nuer in essentialist terms. In what follows, I adopt a two-layered approach, which explains the conflict between the Anywaa and the Nuer as an interplay between the realities of the local economy (pastoral expansion) and the expansion of the modern state: both of these have a direct bearing on the identification process and changing affiliations. In order to support my position with evidence, I use an event-diagnostic approach, and present two recent instances of conflict between Anywaa and Nuer that reveal the issues at stake.

Case study 1: Anywaa-Nuer conflict, Itang, 1998

In January 1998, there were a series of clashes between Anywaa and Nuer villagers in the Itang district of Gambela, which lasted for about six months. What started as a family affair engulfed all the Nuer and Anywaa who lived between Jikaw and Itang districts. The crisis threatened to escalate into an all-out regional confrontation

---

5. Gambela town is divided into ethnic enclaves, particularly the Nuer forming a distinct settlement. Nearly all Nuer residents of Gambela town live in a place called 'New Land'. Highlanders, also distinguished along ethnic lines – Amhara, Tigreans, Oromo and Kembata – all have their respective neighbourhoods.

between the Anywaa and the Nuer. Many villages on both sides were burned down, hundreds of people were displaced and many lost their lives.

The immediate cause of the conflict was an inter-ethnic divorce case. A Nuer from the Cieng Nyajani section of the Gaajak Nuer, a small settlement in Itang away from their major settlement area in Jikaw, asked for the return of bridewealth after a divorce. The Anywaa family refused or was unable to pay it back. In response the Nuer killed his father-in-law and this ignited the conflict between the two communities. By January 1998 the inter-ethnic tension in Itang district had escalated into a major confrontation between the two communities. For a better understanding of what had happened, I discuss here the various actors involved in the conflict. In order to follow this discussion more easily, some of the major names and Nuer categories that are in one way or another involved in the identity politics are listed in Table 8.1.

Table 8.1 Major Nuer names and categories involved in Gambela identity politics

Jikany	Generic term for groups of Nuer who inhabit both sides of the Ethio-Sudanese border
Gaajak	Branch of the Jikany, consisting of the Cieng Chany, Cieng Wau and Cieng Nyajani
Gaajok	Branch of the Jikany
Gaagwang	Branch of the Jikany
Thiang	Branch of the Jikany genealogically but allied with and considered to be Gaajak
Cieng Reng	Branch of the Jikany with a separate origin but allied with the Gaajak
Makot	Cieng Reng settlement in Itang district
Cieng Dung	Section of Thiang that are in competition with the Cieng Reng at Makot

The Cieng Reng community

The Cieng Reng are a section of the Nuer who live mainly in a place called Yom in the Sudanese Jikaw district close to the border. They have a close relationship with the Gaajak Nuer, who live on both sides of the border and in whose wars they have been involved at different levels and at different times. In fact, groups of Cieng Reng have already started living on the Ethiopian side and have 'become' Gaajak or are considered to have done so by other Nuer. The Cieng Reng also draw on their mythical connection to the Gaajak, particularly the Thiang, in whose area they settled.[6] In the

---

6. The Cieng Reng do not belong to the Jikany Nuer by descent, but nevertheless have close historical and social ties with them, particularly with the Gaajak. The mythical founders of Cieng Reng (Tik) and Jikany (Kir) are believed to have favoured each other reciprocally, which later on laid the foundation for profound social ties between the two communities. Tik, master of the water, helped Kir and his followers cross a river during their migration whereas Kir paid the bridewealth for Tik.

early 1980s, a small section of Cieng Reng, initially only a few families but now estimated to be around 2,000 people, came to the Itang area and settled at a place called Makot/Puolkot (referring to the wet- and dry-season settlement sites, respectively). Various reasons have been put forward as pull factors for their settlement: access to refugee resources; access to dry-season pasture and riverain cultivation; the civil war in southern Sudan and the prevailing insecurity for pastoral mobility; access to services, particularly market outlets to the neighbouring Itang town; and competition over local leadership. Kong Diu, the leader of the Makot people, as they came to be known, is from a small group of the Cieng Reng community in Jikaw district. He was born and grew up on the Ethiopian side of the border, but later lived with the wider Cieng Reng community in Yom, southern Sudan. He was already well acquainted with the Itang area, which his people had previously used as dry-season pasture. Having failed to secure a leadership position in Yom, he led a small group of Cieng Reng from Yom to the Itang area and settled at Makot/Puolkot in 1984. Through this he linked his own personal ambition with the perennial desire of the Cieng Reng community to have a foothold along the River Baro/Sobat.[7] A homicide and the resultant feud between two groups of Cieng Reng communities in Yom brought a new group of immigrants to swell the ranks of Kong Diu. With the intensification of the civil war in southern Sudan in the 1990s, the Cieng Reng settlement at Makot increased tremendously and emerged as the biggest Cieng Reng community in Ethiopia. The Makot people have become very prosperous, combining their pastoral economy with an increased cultivation of rain-fed and riverain land, as well as developing new market outlets in Itang and Gambela towns. Over time, they managed to create links with the neighbouring Anywaa communities, and were able to obtain access to land for the valued flood cultivation, through intermarriage, gift exchanges and payments. The Anywaa did not feel their presence as a threat as it did not result in any displacement; the settlement site was originally forest-land. However, in 1997 things changed, when the Makot settlement became a subject of public debate because of a local election and the resultant politics of recognition and identification. Goaded by party-politicking framed in ethnic terms, Cieng Reng, on their part, felt the need to achieve political recognition to protect their cattle from theft by other Nuer and local Anywaa, and also to give them access to relief goods. In the local parlance, gaining recognition meant asking for *kebele* status (the smallest political and administrative unit). On that basis, the Makot people asked for their own *kebele*. This very much alerted Anywaa political actors in Itang district, as it threatened their political domination of the district.

The Thiang Nuer community
The Thiang are one of the main divisions within the Jikany Nuer. After a series of conflicts in the early phase of the Nuer eastward migration in the late nineteenth century, they managed to establish closer links with the Anywaa than any other Nuer

---

7. The rivalry over local leadership in Yom was between Kong Diu and Wengbot and it was also manifested in the religious field: both employed Dinka magicians. Whereas Wengbot bought one of the most powerful Dinka magicians, one of Kong Diu's wives is also a well-known Dinka magician.

section. They now claim all the rangeland between Itang and Jikaw, which they call *nyam duar*, their hinterland. This is an assertion of ownership rights over a key resource (grazing land) through an ideological and political use of genealogical seniority. In the Jikany origin myth, Thiang is the eldest son of Kir, the founding father. The Thiang consider themselves to be a frontier community and have a paternalistic discourse about their relationship to other sections of the Jikany Nuer.[8] Perhaps more importantly, the Thiang also claim seniority as the first group of Nuer who settled in the Jikaw-Itang area and also in Ethiopia at large, bordering the Anywaa.

The Thiang Nuer and the Opeeno Anywaa (Anywaa who live along the River Baro) seem to have developed a subtle alliance. The Anywaa have come to regard the Thiang as a buffer zone sheltering them from less familiar and often more aggressive Nuer sections, while the Thiang have come to consider all unoccupied Anywaa land as their hinterland. The Thiang are generally not happy for other Nuer to by-pass them and deal directly with the Anywaa. They prefer the newly arrived Nuer to attach themselves to a nearby Thiang settlement and thereby contribute to their group strength. Thus the Thiang consider the presence of other Nuer in Itang area as a destabilizing factor. They have a vested interest in the status quo and see themselves as strategically vulnerable in the strained relationship between the Anywaa and Nuer.

Against such a background, a specific section of the Thiang, called Cieng Dung, has become extremely hostile to the Makot people (Cieng Reng), whom they envy because of their tremendous cattle wealth and access to the Anywaa's riverain land.[9] A group of Cieng Dung families was brought by the Anywaa to settle in areas claimed by the Makot people, in response to a request by the Cieng Dung. On their part, the Anywaa gave support to the Cieng Dung/Thiang, because of their long-standing contacts and also to counterpoise the steadily growing and assertive Cieng Reng community at Makot. Having securely accessed the riverain lands, the Cieng Dung adopted the same exclusionary discourse (the 'citizen card') in an increasingly vocal way in their land dispute with the Makot people. Like the Anywaa, the Cieng Dung labelled the Makot 'foreigners'. In the conflict that ensued between the Anywaa and the Nuer in 1998, the Cieng Dung and some sections of the Thiang joined the Anywaa, while other sections of the Thiang joined the Makot people.

Anywaa political actors in Itang district
Itang district is one of the most politically contentious areas between Anywaa and Nuer in Gambela region. The number of Anywaa and Nuer are about the same, but the *kebele* distribution is very disproportionate, as Anywaa are given fourteen of the

8. The vanguard position of the Thiang seems to have been acknowledged by Ngundeng, the greatest of all Nuer prophets in the late nineteenth century. Ngundeng is said to have sanctioned the settlement pattern of the Jikany in one of his prophetic songs: 'Let the Thiang go first, and other Jikany follow!'
9. The Cieng Dung are the tertiary division of the Thiang Nuer, which together with the Fangak form the Cieng Chuol. As they are a numerical minority, the Cieng Dung have attached themselves to other Thiang sections as well as allying themselves to the Anywaa.

twenty *kebeles*. They occupy the chairmanship of the district council, as well as heading the security and other key offices. The main basis for the size of political representation is the discourse of autochthony. The Anywaa fervently claim Itang district to be their territory, and consider the Nuer in general to be latecomers. Against this background, when the Makot people asked for a *kebele* status, it was perceived as a threat to the political status of the Anywaa. The issue escalated, as the Makot people asked not only for one *kebele* but for six *kebeles* in proportion to their population size. The most effective strategy for promoting the exclusion of the Cieng Reng was for the Anywaa to label the people of Makot 'refugees' and hence foreigners. With the coincidence of political interest between the Thiang/Cieng Dung Nuer and the Anywaa political actors, the stage was set for confrontation between the various interest groups.

Nuer political actors in the Gambela regional council
The Nuer political actors in the regional council took the Makot case seriously. A number of Nuer officials visited the area and lobbied for their recognition as a *kebele*. Such a high political profile ensured that the issue became further ethnicized. The same exclusionary political discourse was being used by Anywaa political actors about the Nuer officials and civil servants in the regional power game and in the job market. Many of the Nuer officials and nearly all of the Nuer civil servants are said to have been educated as southern Sudanese in the refugee camps within Ethiopia.[10] What was social capital in the 1980s (switching from an Ethiopian citizenship to a more rewarding refugee status by activating Sudanese connections) became a political liability in the 1990s when the incentives were to 'become' Ethiopian because of the expansion of the state in the post 1991 period and the new opportunities that came with it. As a result, Nuer political actors in the regional council were determined not to set a precedent by letting the Anywaa expel the Cieng Reng from Makot village.

The politics of entitlement
In 1998, a small incident that had begun as a family affair (an inter-ethnic divorce case) became a major ethnic conflict. In Itang more than eleven villages were burned down and many people lost their lives. Kong Diu and other 'ringleaders' were imprisoned for a year, and attempts were made on his life and to deport his people. The Makot people not only were denied access to riverain cultivation, which they had managed to obtain through various forms of social networking and payments to the local Anywaa, but also were prohibited from using the River Baro.

The politics of recognition, however, has continued to this day. When the issue entered a deadlock at the regional level, it moved to the federal government. While the Anywaa-dominated regional council[11] produced a document to prove the 'non-citizenry' status of the Makot people and the potential danger of granting them

---

10. These include senior Nuer officials in the regional council, ambassadors and bureau heads.
11. The Anywaa have occupied the office of the regional presidency since 1991 and filled key ministerial offices, such as security, planning, relief and development, as well as 45 per cent of the seats in the council, whereas 37 per cent of the seats are allocated to the Nuer.

recognition, Kong Diu himself travelled all the way from Makot village to Addis Ababa, the national capital, to appeal to the federal parliament.

One of the main strategies used by the Nuer political actors and members of the Cieng Reng community in their politics of recognition is a 'historical' argument. They contest the Anywaa claim to autochthony by emphasizing their common origin in southern Sudan and the history of Nilotic migrations. To enhance further their relatedness to the Ethiopian state, the Nuer also activated what they consider to be the reciprocal relationship between Ngundeng, the nineteenth-century Nuer prophet, and Emperor Haile Selassie. The myth of reciprocity between the Nuer and the Ethiopian state is based on the tale of a tail: the story of an ox and other gifts that Ngundeng is believed to have given to Emperor Haile Selassie in his war against the Italians during the Second World War. The tale is superimposed on earlier stories of encounters between the forces of Emperor Menelik who were sent to the Upper Nile region in support of the French in their colonial competition with the British. In what appears to be the telescoping of four decades of time, between Emperor Menelik and Emperor Haile Selassie, history is compressed by the Nuer to provide subjective meaning and to create political capital in order to counter the Anywaa's politics of exclusion. The references to 'history' and the attendant moral discourse serve the purpose of familiarizing oneself in a contested area and help to make contact with and make demands on the state to be responsive to one's plight.

Another Nuer entitlement discourse bestows ownership rights to land and people upon the state (*kume*). The Nuer often say, 'Land, water and all the other resources belong to the state.' This is more of a local statement used as an exit strategy to counter an exclusive discourse of autochthony by the Anywaa than a commitment to the state or an assertion of national identity. It is a way in which local actors manipulate creatively an ideology of the state in the contestation over political power and natural resources.[12]

The more informed segments of the Nuer population also refer to the modern state concept of 'naturalization' in the politics of recognition. In their perspective, eighteen years of stay (the first Cieng Reng settlement at Makot started in 1984) should suffice to entitle citizenship. These claims have helped the Cieng Reng frame their politics of recognition in the language of the state.[13] Recognition is also couched in the political jargon of having paid tax to the state. The Makot people claim citizenship on the basis that they paid tax to the state when they were part of an Anywaa *kebele*, prior to their request for their own *kebele*. In this politics of entitlement different logics concerning identity and belonging can be seen to be at work. The Nuer project their own indigenous political system onto the national

---

12. All the regimes, the imperial, the socialist and the current federal state, have controlled land in the name of the people.
13. Ethiopia still does not have a refugee law or policy. During the imperial and the early phase of the Derg, the trend is towards a more integrative approach between refugees and the local people. Such was the case for the group of Nuer, Anywaa and Shilluk from Anyanya-I, who were absorbed by their respective communities. The time frame given for naturalization was three years.

state, as if it is nothing but Nuer writ large. Accordingly, 'the Ethiopian state should celebrate when new people join it', since politics is viewed by the Nuer as essentially the politics of numbers: the bigger you are, the stronger you become. The most informed segment of the community, including Kong Diu himself and students from Makot, give the issue a 'global' touch as well. While elaborating on the issue, Kong said, 'Why should the Anywaa prevent us being Ethiopian, while everybody, themselves included, go to other places such as America and become Americans?' This is a reference to the steadily increasing number of Nuer, Anywaa and other Ethiopians who have settled in the United States.

The outbreak of the border war between Ethiopia and Eritrea in May 1998 has been a further catalyst in these local politics of recognition. Many people from Makot were recruited into the Ethiopian army and the issue of entitlement has not only become a moral question but also assumed a political dimension. In the words of Kong Diu, 'The Cieng Reng will go back to Sudan only if the Ethiopian government brings our people back from the battle front, including the dead ones. We are now as Ethiopian as the Anywaa are!'[14] In order not to antagonize the Anywaa-dominated regional state of Gambela the federal government responded to the Cieng Reng's politics of recognition half-heartedly. Although they were not recognized as a *kebele* they were given a 'residence permit' in Makot village.

Case study 2: battling for the future: conflict in the schools

Between 1997 and 1999, there were a series of clashes between Anywaa and Nuer students in Gambela town. The first students' clash occurred in 1997. It was triggered by a controversy over whether Nuer students should be allowed to join the TTI (Teacher Training Institute), the only institution of higher learning operating in Gambela region.[15] As in other cases, many of these students are also 'ex-refugees' who completed their secondary school education in the southern Sudanese refugee camps in Gambela. Many Ethiopian Nuer had joined the refugee camps in order to gain access to education provided by the UNHCR (United Nations High Commissioner for Refugees) and other aid agencies. Very few Anywaa had managed to do the same, as they were conspicuous as Ethiopian nationals and regarded as such by the aid agencies. This, of course, created resentment among the Anywaa, who were later able to turn their predicament into political capital. The Nuer who were educated as

---

14. The war has created new national-cum-local heroes, who are creatively used as political capital in the identity game. Bill Bichok is one of such heroes. Bill is one of the Nuer from Gambela region who joined the Ethiopian army when the Ethio-Eritrean war broke out in May 1998. In one of the military engagements Bill is said to have performed a 'heroic' act of repulsing advancing Eritrean forces. He was later on decorated as one of the 'Badime heroes'. The ruling EPRDF party produced a poster in memory of the 'war of sovereignty', which featured a picture of Bill. When the poster reached Gambela, it became an important argument used by the Nuer to connect themselves with and make demands on the state, which was very important in the intense politics of exclusion from national identity and the regional power game they have been engaged in with their neighbours, the Anywaa.

15. The institute is now upgraded to a College of Education and Nursing.

refugees found themselves vulnerable to the politics of exclusion in the 1990s, as they had little access to political power and the job market. This was the background when Anywaa officials in the Bureau of Education rejected the applications of groups of Nuer students to join the TTI. The rejection was on the grounds that they were Sudanese, not Ethiopian citizens. They were considered to be refugees 'who don't even know Amharic', one criterion in defining who is and is not an Ethiopian. Later on, it was revealed that Anywaa officials allowed, in fact encouraged, Anywaa students and teachers from Sudan (Khartoum) to join the TTI, arguing that there was a shortage of local people to implement the new educational policy.[16] As the Nuer students carry their 'Sudanese identity' in their transcript and are familiar from refugee camps in Ethiopia, they became an easy target for exclusion. The controversy spread later to the schools in Gambela town and led to a confrontation between Anywaa and Nuer students. Many students on both sides were stoned, and around five Nuer women who had been fetching firewood were found dead in the nearby forest. After intensive political bickering between Anywaa and Nuer officials in the Bureau of Education and the Regional Council, the Nuer students were allowed to study at the TTI.

Two years later, in November 1999, there was another violent clash in the schools. This time it was caused by yet another controversial issue: which language to promote in the new education policy. Gambela is one of the four regional states in the Federal Democratic Republic of Ethiopia that adopted Amharic as the working language of government. Competence in Amharic is still a prerequisite for employment in political and administrative offices. The new education policy, however, draws on the 'mother tongue' approach. Accordingly, the curriculum has been changed from Amharic into 'majority' local languages. It was therefore decided to start the new curriculum in Anywaa and Nuer languages in 1996 in the Gambela regional state. By then there were more Anywaa than Nuer teachers. So-called 'tryouts' were initiated in the region, three in Anywaa areas and one in a Nuer area. While the tryout could be started immediately in the Anywaa areas, the teachers in the Nuer area were all highlanders. This created a differential start. By 1999 the Anywaa had implemented grades 1 to 8 in all schools, while the Nuer were just in the tryout phase of grade 8. In Gambela town two elementary schools were allocated to Nuer while the rest were allocated to Anywaa. Upon completion of their elementary school, Nuer students from Gambela town had to join junior secondary school, a school where the curriculum was designed in the Anywaa language. In the first cycle (grades 1–4), the medium of instruction is the Anywaa language and Amharic is taught as a subject from grade 3. In the second cycle (grades 5–8), the medium of instruction is English and Anywaa is taught as a subject. This arrangement created resentment among the Nuer. The Nuer who went to this school would have to learn Anywaa as a subject. The result was that Nuer students, as well as students from other ethnic groups, felt deprived and disadvantaged. The Nuer considered the issue to be an attempt to make Anywaa the regional language, and

---

16. Bureau of Education, letter dated 21 October 1996, File No.#138/13/6.

hence to further their own marginalization. The contentious issues were not only exclusion from and differential access to the means of social mobility, but that ethnic honour was at stake via language and its connection to identity. Extreme elements on both sides advocated the promotion of their own language as the language of the regional government. This issue was played out in another question: who 'owns' the regional capital, Gambela town? The Anywaa claim that it is theirs and until recently they were the majority. According to the 1994 census, the highlanders form a majority in Gambela town (55 per cent), followed by the Anywaa (34 per cent), and the Nuer (10 per cent). However, since the census, the number of Nuer and highlander residents in Gambela town seems to be steadily growing. This has given the town a more multi-ethnic face.

The crisis is said to have been precipitated by the appointment of a Nuer as head of the curriculum department. Tension started when Nuer students boycotted Anywaa lessons and established a 'Nuer Student Union', demanding 'equal opportunity'. Students grouped along ethnic lines and started throwing stones at each other. In a desperate attempt to de-escalate the conflict, police used excessive force, killing one Anywaa student and wounding many from both groups. The situation was aggravated by the 'rumour machine': it was rumoured that the Nuer were coming from Jikaw district to take over Gambela town, and the Anywaa were said to have asked for reinforcements from the neighbouring Abobo district to 'storm' Newland, the exclusively Nuer neighbourhood in Gambela town.[17] As the political crisis escalated, people were confined for weeks in exclusive ethnic no-go zones: the Jejebe Bridge across the River Baro symbolically and physically divided the two communities. Past conflicts between the two communities were invoked and traditional war songs were chanted to inject a sense of continuity with past enmities.[18]

The crisis abated after a general public meeting was convened to look into the root causes of the problem. After heated debate, it was decided to teach in both languages. But perhaps the most important feature of this meeting was the chance for each group to articulate its group interests. The Nuer used the forum to voice their political grievances. They protested about what they called Anywaa domination of the regional state, their occupation of all the key offices and the way in which this is supposed to have created an imbalance in resource allocation. On their part, the Anywaa raised the question of Nuer expansion and expressed their anxiety.

---

17. Christiane Falge, personal communication. She was an eyewitness of the conflict.
18. In recent days there is often a reference both by Anywaa and Nuer to Tieragak, one of the major conflicts between the two communities in the early twentieth century. The battle took place in Edeni, Jikaw district, between the Opeeno and the Jor Anywaa on the one side against all sections of eastern Gaajak Nuer on both sides of the border. But it is remembered as the battle of Tieragak, the furthest Anywaa village from Nuer, near Bonga town, which the Nuer are said to have pledged to push the Anywaa to. War songs from this conflict are popular and often invoked in contemporary confrontations. It is common to hear Anywaa saying 'the Nuer will not stop until they reach Tieragak'.

Analysis

The literature on collective identity has long been dominated by the sterile debate between the so-called primordialists and constructivists. The problem with the primordialist approach is their insistence that ethnic groups are 'fixed entities', which makes it impossible to 'explain why some ethnic groups decay, new ones appear and others merge. They cannot tell us why some characteristics seem more important than others and why some ethnic groups, seemingly as a whole, fight each other and others cooperate' (Wieland 2001: 210). Perhaps more importantly, primordialists tend to imply that primordial feelings lead to ethnic conflict. However, the primordial position reminds us of the importance of connections conceived in terms of descent: notions of origin and relatedness can be put to political use. At the other end of the theoretical spectrum, the constructivists, particularly the instrumentalist variety, subscribe to an extreme form of subjectivism. They tend to neglect the factor of origin, the resilience of a speech community and 'culture areas'/ethnic categories. They often fail to distinguish ethnic groups from other cultural categories. Above all, they have difficulties in explaining why the masses tend to be mobilized so easily through appeals to origin and culture, and even why people are ready to die for such invented, constructed categories (Wieland 2001: 210). If the constant cannot explain the variable (the primordial position), the variable just explains away the constant (the constructivist position). The real challenge is to demonstrate historically and ethnographically the reification process, instead of assuming that it is 'ineffable', and to show how an 'uncaptured', revived, imposed or constructed/invented social category is charged with subjective meaning and is made relevant to everyday life (Thompson 1989).

Following the new 'third way' (Scott 1990), I propose a two-layered approach to explain the conflict between the Anywaa and the Nuer and the political debate between the two: 'conflict from below' (pastoral expansion) and 'conflict from above' (the hegemonic project of the state and the ensuing politics of exclusion). It is the interplay between these two forces that has resulted in the intensification of 'ethnic' conflict in the Gambela region. In the historical context of pastoral expansion, identity politics is fought with a repertoire of collective memories of winners and losers. The experience of conflict is also predicated on cultural practices related to differing discourses of identity (the 'primordial' dimension), whereas the expansion of the state becomes a catalyst in the current ethnification process (the constructivist dimension). In the context of the identity politics between the Anywaa and the Nuer in Gambela region, collective memory puts a limit on agency and on the extent to which local political actors are able to manipulate group identity in the competition for power (Schlee 2000a). Individuals operate within a system of constraints (the resilience of the public discourse and the sets of historical experiences) and yet also respond to new opportunities and rewards in the new game of identity politics offered by the expansion of the state, which has activated and made these memories relevant.

The Question of Nuer Expansion and the Politicization of Memory: Conflict from Below

One of the main academic debates in the anthropological literature of the 1970s and 1980s concerned the dramatic eastward expansion of the Nuer. Four models have been presented to account for the phenomenon: demographic (population pressure – Evans-Pritchard 1940b; Sahlins 1961; Hutchinson 1996); ecological (the expediencies of flooding and the resultant population movement – Johnson 1986); social (cultural practices, particularly high bridewealth payments – Kelly 1985); and superior military efficiency (Otterbein 1995). Whatever the reason(s) for the expansion and success of the Nuer, by the end of the nineteenth century the Nuer had increased their territorial domain fourfold, displacing various communities, particularly the Dinka and the Anywaa (Kelly 1985). Since the military conquest, the Nuer expansion has assumed a largely 'peaceful' form through intermarriage and various means of assimilation.

If Nuer expansion was an academic debate decades ago, it is now a political issue in inter-ethnic relations in Gambela. The memories of past losses are being invoked by Anywaa political actors but certain new developments are being perceived of as signs of a new eastwards expansion that has the 'grand design' to take over all the major rivers of Gambela that fall within Anywaa areas. These new developments include: the influx of Nuer refugees; the emergence of new Nuer settlements near the major towns and amidst Anywaa territory; and the growing visibility of Nuer in the regional politics. The resultant debate on the relevance of the past and its current interpretation requires an approach that goes beyond an examination of the military expansion of the Nuer at the expense of their neighbours.

One area for consideration is the social ethos of the groups under question. There is a marked difference between Anywaa and Nuer discourses of identity. There is a strong sense of territoriality and exclusivity among the Anywaa. The notion of a territory proper to each form of existence and the absolute right of each part to a peaceful existence in its own territory is fundamental in an Anywaa's conception of the structure of the universe. Accordingly, all have the right to exist, but only within the limits of their territory (Perner 1994). Such a cosmology governs not only the relationship between the domains of the earth (the human territory) and the sky (the world of spirituality) but also applies to inter-ethnic relations. Current Anywaa identity discourse also gives prime importance to descent. They consider that, to be fully Anywaa, the patriline of the mother should also be Anywaa. Purity of descent is, therefore, an important feature in Anywaa identity discourse. Both the father and the mother lines are important in clan greetings (*math*) and titles (*paak*). Each man and woman has his/her *paak*. A woman takes up the title of her father's clan, while a man that of his mother's (Kurimoto 2000: 3–4). Children born from inter-ethnic marriages often bear within their names connotations of their 'foreign' background. *Ugaala* and *Nyigala* are typical Anywaa names for those born from a highland father or mother, denoting that he/she has become *gaala*, highlander. In contrast, the Nuer give a name to people with foreign or mixed background that denotes their becoming Nuer. *Nyanuer* is one common name, 'she became a Nuer'. *Nyanuer* means

not only to become Nuer, but also to behave in the Nuer way. It is contrasted with
Nyaduar, which means being brought from the wilderness: a subtle reference to
people who have other ethnic backgrounds. By stressing the origin in the
'wilderness', the moral and social superiority of being Nuer is implied.

On the other hand, some Anywaa still trace their descent from Nuer or Dinka,
suggesting an earlier more inclusive discourse of identity. The 'primordialist' turn in
Anywaa identity discourse might have been defensive, as seems to have been the case
among the Dinka, who 'reaffirmed the fundamental indissolvability of their ethnic
identity through an elaboration of blood-based metaphors of procreative descent
against the sticky grasp of their Nuer neighbours' (Hutchinson 2000: 12). Against
such a background, the Anywaa found the encroachment of their territory by their
neighbours incomprehensible or, worse, a 'plot' to threaten their existence.

In contrast, the Nuer social and political system is strongly assimilationist. The
Nuer adopt large numbers of captured children and even adults. Nuer identity is
basically defined as performative (Hutchinson 1996). To be Nuer is to behave in a
morally appropriate way, which is often contrasted with other groups' behaviour.
Until recently, to be a Nuer man, called *wut*, was also to bear the six horizontal scars
on the forehead, called *gar*, not only celebrating manhood but also as a 'lifelong
commitment' to being Nuer.[19] To be Nuer is also equated with love for cattle. By
definition, every person could become Nuer as long as he/she speaks Nuer and
exhibits the appropriate behaviour.[20] Another interesting means of effective
assimilation is giving community leadership positions to outsiders (Hutchinson
1996). There are many instances when an Anywaa minority leads a Nuer group in
its war against other Nuer group. Another typical assimilationist mechanism is
territorial encirclement. This creates a sense of insecurity on the side of the Anywaa,
and often they either leave the area for a new place or become gradually drawn into
the Nuer orbit. In the latter case, they become cut off from the wider Anywaa unit,
and the assimilation that results is not by choice but a fait accompli.

As discussed above, the Nuer political system also draws on 'the politics of
numbers': the bigger you are, the stronger you become. Accordingly, outsiders are
not enemies to be avoided but objects of assimilation. A diversified livelihood
strategy has given the Nuer an economic clout that facilitates the process of
assimilation, not only creating the image of the 'richer' Nuer, but also to enable them
(individuals and groups) to give favours to make their neighbours indebted to them
and create social and interpersonal networks as well. Above all, it has created

19. With the spread of education and increased contact with other groups, the significance of *gar*
    has declined. A growing number of Nuer schoolboys and those who are born and growing up
    in the towns don't have *gar*, at times pejoratively called *tut dual*, bull boys (grown-up boys).
20. In practical terms, Nuer put a limit on assimilation. With more encounters with new groups
    of people, and other discourses of identity, they have become more aware of the 'physical'
    boundary. This is especially true for the 'red' people, the highlanders. Nevertheless, there are
    instances of assimilated highlanders during the resettlement period. Some highlander women
    and children were taken as captives when they were escaping the resettlement sites in Gambela
    to Sudan.

asymmetrical marriage relations between the Nuer and most of their neighbours. Hence, Nuer men marry Anywaa women, a one-way traffic.[21] The reverse is the exception, which proves the general rule. Inter-ethnic marriages are more pronounced and politically motivated in areas of new settlement. There is an economic rationality to assimilating others. Recruiting new members could be seen as an economic strategy to exploit wider ecological niches. As the population expands demographically, territorial expansion becomes imperative. Nuer livelihood strategy is very much diversified and is based on extensive mobility, moving to milk-, grain- and fish-producing areas seasonally (Evans-Pritchard 1940a).

In the village setting, new Nuer settlements are being established near Anywaa villages along the banks of the Rivers Baro (e.g. Pinkeo, Abol), Akobo and Gilo in order to secure dry-season pasture as well as access to riverain cultivation. The banks of the River Baro 'are considered as some of the richest and most coveted agricultural lands in all of Nuerland because they yield lush crops of maize and tobacco during the height of the dry season with only the moisture that rises from the river' (Hutchinson 1996: 114). The flooding pattern of the Baro also provides the best conditions for after-flood cultivation, as it replenishes the ground with fertile alluvial soil. A successful Nuer is one who keeps cattle, owns rain-fed farmland and accesses riverain cultivation, most of which falls within Anywaa territory. The ideal is to attain the status of *ram mi diel*: big family, large herd and vast land with convertible social, economic and political capitals. Polygamous families wield high social capital that could readily be converted into economic capital, as the labour process determines production size. On the other hand, larger herd size (economic capital) could be translated into political capital through exchanges, feasting and inter-ethnic marriages, which in turn bring more economic capital and access to riverain lands. According to a micro-census I conducted in Pinkeo village, 60 per cent of the seventy-six Nuer households were polygamous. This sharply contrasted with the much lower figure of 16 per cent of the sixty Anywaa households.

Most of the current Nuer expansions take a peaceful form, pioneered by individual families settling along the rivers, through marriage relations and personal networks, while in some areas the displacement of Anywaa villagers because of various forms of conflict has facilitated the settlement of groups of Nuer in the vacated areas. The original settlers often attract new immigrants, forming satellite Nuer villages amidst Anywaa settlements.

In the context of the civil war in Sudan pastoral mobility took a more easterly direction towards the Gambela region, causing an influx of Nuer groups en masse, adding not only to Anywaa anxiety but also to friction with earlier Nuer immigrants and settled communities. Such demographic trends seem to have induced the Anywaa discourse of ethnic extinction and the political project of containment.

---

21. The average bridewealth payment for the Nuer is twenty-five cows, while for the Anywaa it is 1,500 birr. After the forceful abolition of the payment of beads, called *dimuy*, by the Derg, bridewealth was monetized among the Anywaa. If a Nuer marries an Anywaa, he pays either 1,500 birr or its equivalent of three cows.

The situation has been compounded by uneven distribution of modern services (health, education, water supply schemes, market outlets), with an 'enclave development' centred on the towns in a peripheral and marginalized region such as Gambela. Service distribution seems to have been dictated by proximity to the towns and to have had a 'roadside bias'. Anywaa areas are better served, as they are closer to the towns. Such an imbalance is nevertheless increasingly perceived in ethnic terms. The Nuer bitterly complain about the fit between political representation and access to public goods and services. New Nuer settlements are often justified in terms of a desperate attempt to access services and facilities, particularly health and education.

Be the reasons for Nuer expansion as they may, it has created a social context within which the contemporary Anywaa identity politics is being played out. The fact on the ground is the ever increasing expansion of Nuer in the villages as well as in the modern institutions. Analytically, it is a logical outcome from a socio-political system that is geared towards expansion, as well as the growing relevance of groups' strength in the competition for modern goods; for the Nuer it is an urge to 'catch up' in the social competition, while for the Anywaa it is seen as a concerted action to impose Nuer hegemony at best, extinction of Anywaa society at worst.

The Role of the State and its Differential Impact: Conflict from Above

Recent studies have characterized the state in Africa as 'collapsed', 'failed', 'weak' or 'inverted' in face of marginalizing global forces (Chabal and Daloz 1999). Contrary to the current emphasis on the decline and the 'irrelevance' of the state in Africa, I argue that the state still matters, not only as an embodiment of 'sovereignty' but also as a force that strongly affects the local political processes. Despite its impoverishment in absolute terms, the state is still the main provider of jobs, and commands a hegemonic position in ideas of belonging. In Ethiopia, state power is also embedded in its control over land.[22] The mere presence of the state brings a new 'actor' into the process of defining and redefining ethnic boundaries. Ethnic groups delimit themselves not only vis-à-vis neighbouring groups, but also vis-à-vis the state in which they are living. This is particularly true in the case of the Ethiopian state, which has built up its structure through an ethnic framework and whose activities have impacted differentially on the various social groupings that live within its domain.

Ever since the expansion of the imperial Ethiopian state in the late nineteenth century to the Gambela area, Anywaa society has been adversely affected by state encroachments. There were attempts to co-opt the local elites during the second phase of the imperial period (1941–74), but the Anywaa were largely excluded from the power game and, when they attempted to resist external control, they faced repression. By and large, however, imperial rule was tenuous, except in the commercial centre, which was jointly administered with the British colonial forces (Bahru 1976).[23]

22. The Ethiopian state, across regimes, has consistently claimed the ultimate ownership of land. Even at the heyday of socialist rhetoric of 'land to the tiller', the praxis was 'land for the state'. The current ethnic regime still claims land in a typical neo-patrimonial discourse.
23. According to the 1902 boundary agreement between imperial Ethiopia and the Anglo-Egyptian colonial establishment in Sudan, Britain was granted a commercial post in Gambela on the navigable River Baro, 400 hectares on the right bank of the river.

State encroachments were felt much more during the Derg military socialist state (1974–91). The more benign indirect rule of the imperial state was replaced by an aggressive penetration of Anywaa society. In what came to be called *Yebehal Abiyot* (Cultural Revolution Ethiopian style) and its hyper-modernist discourse, Anywaa culture was suddenly and violently uprooted.[24] Hence, their beads, with which they marry, were thrown into the river and banned by law, and their traditional leaders were disgraced and deposed.[25] Local resistance to such impositions were violently suppressed in a series of military engagements with the local leaders. In the then dominant Marxist discourse, these were seen as conflicts between the *adhari* (reactionaries) and *teramaj* (revolutionaries).[26] The sudden disconnection from cultural values that resulted was not replaced with an alternative value orientation except for a hollow revolutionary rhetoric. This seems to have caused serious social disruption among the Anywaa.

The commencement of the civil war in southern Sudan and the attendant proxy war between the government of Sudan and the Ethiopian government had a tremendous impact on local politics and the identification process. Refugee camps were established in Gambela, which hosted about 300,000 people by the mid 1980s. The rebels, particularly the SPLA (Sudan People's Liberation Army), also used these camps for recruitment and as training grounds. The SPLA was heavily supported by the Derg, which it reciprocated by becoming involved in the military actions of the Derg. These new political developments affected local politics in two ways. First, it brought insecurity to the region and military clashes between the refugees, who were predominantly Nuer, and the local Anywaa. At times this spilled over to the local Anywaa and Nuer villages. Secondly, it brought disruption to the local economy: the new relief regime and refugee economy discouraged local production. There was an unprecedented fall in the market price for local grain. A quintal of maize dropped from fifty birr to four birr, and reports tell that the sack that contained the maize was more valuable than the grain itself (Kurimoto 1996). The net result of the 'revolutionary experience' of the Anywaa is, therefore, instability and intensification of inter-ethnic conflict, the spread of negative symbols of modernity (alcoholism, outmigration from the villages and rapid urbanization, diseases and prostitution). A continuing loss of confidence because of material deprivation resulted in a rapidly declining group position in the local power relations (Kurimoto 2000). An ill-fated

---

24. The military socialist state of the Derg adopted the ideas of 'cultural revolution' from the Chinese model, which had a huge influence in Marxist discourse and prevailed particularly during the formative stage of the revolution (Solomon 1993).

25. Evans-Pritchard referred to the Anywaa as the 'beads people', contrasting them with their pastoral neighbours, who use cattle for bridewealth payments. The most important Anywaa beads for bridewealth are the blue-coloured necklaces called *dimuy*. After the forceful confiscation of *dimuy* bridewealth payment was converted into cash. In remote parts of Anywaa (Jor) and the Adongo/Pochala areas of southern Sudan, *dimuy* is still used.

26. The resistance was consolidated in Jor district, where the local Anywaa leaders managed to contain the abolition of the traditional political system. It took more than five years for the local Derg authorities to finally subdue the resistance.

and politically motivated resettlement, which brought an additional 60,000 farmers from northern and southern Ethiopian highlands to the Gambela region, further fuelled the inter-group tension. All the resettlement sites were in Anywaa territory, and, in the face of massive refugee influx, the Anywaa found themselves, suddenly, a demographic minority. Not unexpectedly, this induced resentment.

The story is somewhat different with the Nuer. The Nuer seem to have 'benefited' from neglect. Unlike the Anywaa, they were not labelled as reactionary by the socialist regime. They neither had 'feudal lords' nor kept 'superstitious' beads. Besides, they were not within easy reach of the revolutionary state, thanks to their relative distance from the towns. When the Nuer felt the revolutionary state deeply was when the civil war spread into Nuer areas on the Ethiopian side of the border.[27] However, this was short-lived and less consequential than the immense social disruption the Anywaa had undergone. In fact, the new refugee economy enabled a group of Nuer to pursue their personal advancement, mainly in terms of access to better education opportunities in the refugee camps under the UNHCR and other aid agencies. Ethiopian Nuer found it relatively easy to switch to southern Sudanese identity and join the refugee camps, an option that was much more difficult for the Anywaa.

The 'cultural revolution' of the Derg, the increasing lawlessness and repressive posture of the SPLA, the huge influx of immigrants and the forceful conscription of villagers for the war in pre-independent Eritrea (between the Ethiopian army and the EPLF (Eritrean People's Liberation Front) precipitated the establishment of an Anywaa-based liberation struggle, which came to be known as GPLM, Gambela People's Liberation Movement. As the civil war in southern Sudan intensified in the mid-1980s, interstate relations between Ethiopia and Sudan worsened. The Derg heavily supported the SPLA, which to all practical intents and purposes had become the de facto government in Gambela, while the Sudanese authorities reciprocated by supporting GPLM. The GPLM were able to step up their military activities in the camps on the Sudanese side of the border and carry out occasional incursions inside Gambela (Kurimoto 1997). This led to the strategic evaluation of different groups by the Derg. The Nuer were the obvious choice for preference not only because the Anywaa were 'unreliable' after they took up arms, but also because the Nuer could serve as a better link to the SPLA. This created an opportune moment for Nuer political entrepreneurs, who were able to combine their individual ambition with effective service to the state.[28] This provoked the Anywaa, who considered the Nuer to be 'upstarts' and largely 'foreign'. With control over the key administrative and political offices by the Nuer during the Derg, the stage was set for mobilizing ethnicity for political action, a trend that was institutionalized in post-1991 Ethiopia. With better connections with the SPLA and occupying two of the senior

---

27. There were a series of clashes between the SPLA and Nuer civilians in Jikaw district in the second half of the 1980s. These clashes are remembered as the Gaajak-SPLA war.
28. The rise into pre-eminence of Thowat, party secretary of Gambela region during the Derg, was mainly because of his contribution to putting down the rebellion in Jor district and his role in forging effective links between the SPLA and the Derg.

posts in the provincial government, the Nuer became well entrenched in Gambela by the end of the 1980s.

With the overthrow of the militarist socialist regime in 1991 and the seizure of state power by the EPRDF, the multi-ethnic Gambela regional state was created. The new political arrangement brought with it unprecedented financial resources and job opportunities to the region, creating a fertile ground for competition between ethnic elites.[29]

With the consent of the EPRDF, GPLM seized political power in Gambela and established the regional government. The Anywaa have dominated the GPLM, and they are considered to be an organization 'sympathetic' to the EPRDF.[30] The seizure of political power of the Anywaa was preceded by a mass exodus of the Nuer from refugee camps and Gambela town, as the Nuer-Derg officials agitated and promoted fears of an 'ethnic reprisal' by the Anywaa.

The adoption of political ethnicity as the official state ideology since 1991 has been closely related to local political processes and 'ethnic conflict' between the Anywaa and the Nuer. The seizure of power by an Anywaa-based GPLM provoked the establishment of a predominantly Nuer-based party called GPDUP, Gambela People's Democratic Unity Party.[31] The rivalry between the two ethnic parties caused communal violence between Anywaa and Nuer. The ensuing politicization of identity further led to violent clashes between the Anywaa and the highlanders (resettlers in Abobo district), as well as the Nuer and the highlanders in Itang town in the early 1990s. The situation gave the impression of total anarchy and the public referred to it as the time of *girgir*, from the Amharic term for turmoil (Kurimoto 1997).

The political rivalry between the GPLM and the GPDUP was framed in ethnic terms. Anywaa political actors justified their seizure of power through their claims to be the original inhabitants of Gambela, and labelled the Nuer 'refugees'. The new political context provided the 'missing link' between conflict from below and conflict from above. A collective memory (the nineteenth-century displacement of Anywaa

---

29. Gambela Regional State is heavily dependent on the central government, from where it gets more than 90 per cent of its budget. It has not managed to extract local wealth through taxes or to initiate a regional development scheme.

30. The Anywaa have so far occupied the office of the presidency, the Nuer are allocated the vice-president and the Majangir the secretary. The Opo and the Komo are each given a chair in the regional council, whereas the highlanders, except for the large resettler community in Abobo district, which is represented in the *wereda* council, are not formally included in the political process.

31. The founders of GPDUP were barred from entering Gambela by the GPLM, which was fully in control of the region during the transitional period of the new government. Instead, they 'sneaked in' through the neighbouring Dembidolo area, escorted by government forces. Subsequent confrontation between the two parties include the burning of the GPDUP flag in Gambela town and labelling its leaders as foreign nationals, threatening the political stability of the region. The EPRDF finally intervened and imposed a merger between the two parties in August 1998, much in line with establishing satellite organizations in other regions to ease political control.

by the Nuer) was converted into political capital. Anywaa were alerted to the 'ever-growing' Nuer expansion at the expense of their territories. Anywaa politicians evoked emotional responses with phrases such as, 'We were in Khartoum, Nasser was ours'. Such sensationalist appeals seem to have provoked communal violence. When Anywaa killed many Nuer civilians in the 1990s, Nuer burned a number of Anywaa villages in the second half of the 1990s.

Was Gambela ripe for 'ethnic' violence in 1991? Not really. Pre-1991 Anywaa-Nuer conflicts were localized, and there were more reported cases of intra- than inter-ethnic conflicts. Where conflicts began to be ethnicized was in the strained refugee areas (predominantly Nuer) and with local Anywaa. But, even then, there was as much Nuer-on-Nuer violence, related to the activities of the southern Sudanese rebel movements, as clashes between Anywaa and Nuer. The ordinary Anywaa and Nuer in the towns and villages of Jikaw and Itang (the main interactional area) were leading normal lives. As was mentioned in the case study, there was particular integration among the Anywaa who live along the River Baro (Opeeno Anywaa) and the Thiang Nuer, who live in Itang and Jikaw districts. The integration did not follow the typical pattern of asymmetrical relations (Nuer marrying Anywaa, Anywaa 'becoming' Nuer). Thiang also adopted, in the perspective of other Nuer, Anywaa customs and behaviour. This was so much so that other Nuer started considering the Thiang 'Anywaaized Nuer'.

In contrast, Anywaa politicians in the GPLM adopted a different ethnic discourse. Accordingly, 'Nuer is Nuer no matter which and where.' In this regard, Anywaa villagers and Anywaa politicians seem to have adopted different approaches to 'containing' Nuer. With the privilege of hindsight, the villagers' approach is more pragmatist and successful, as the GPLM's indiscriminate categorization helped consolidate Nuer ethnicity. For post-1991 Nuer, ethnicity in Gambela is basically reactive to political exclusion from the regional power game. Interestingly, despite the fact that most of the violent clashes between Anywaa and Nuer have occurred in the areas where the Thiang and the Anywaa live together, the 'Thiang factor' tends to work towards cross-cutting ties than articulating ethnic interests. In fact, a militant Anywaa ethnicity and reactive Nuer ethnicity did not deter the emergence of a separate identity among the Thiang. The new group/ethnic quota system as a means of allocating benefits and state resources, coupled with their strategic vulnerability, led the Thiang to distance themselves from their closest allies, the Gaajak Nuer, in whose name they have widely become known. The new identity game has required a skilful combination of various inclusive and exclusive categories depending on the context and the political gains it entails, ranging from state identities (Ethiopian versus Sudanese), ethnic (Nuer versus Anywaa), Nilotic (contrasted with the highlanders) and clanship (Thiang versus Cieng Reng) to various levels of clan segmentation.

The current political process has tipped the balance of relations between rival groups into ones where ethnicity is the dominant form of political action. Otherwise, there are various forms of identifications at intra-ethnic levels that mimic this discourse of exclusion. The state is a key actor in this identity game, not only setting the rules of the game but also pursuing its own strategic interests. State presence is expanding in the region, as it is located in the border area with Sudan, whose diplomatic relations with Ethiopia are still volatile. Structurally, the new federal arrangement in Ethiopia is a two-tiered system (Young 1999), one that gives more autonomy to the highland states, which are considered to be more 'developed', and the four lowland states, which are considered to be backward or euphemistically referred to as 'emerging regions'. Gambela Regional State is one of these 'emerging regions'. Attached to these regions is the controversial office of political advisers appointed by the Regional Affairs Office under the Prime Minister's Office. Apparently their task is to assist economic development and provide political stability, but they function as de facto 'kingmakers' and as an instrument of political control for the ruling party. They are ethnic Tigreans representing the TPLF, Tigrean People's Liberation Front, the leading organization within the EPRDF. Both Anywaa and Nuer view this office as insensitive, partisan and non-responsive to their respective demands. They perceive that the federal government has so far lacked the political will and the structural capacity to mediate between competing groups and conflicting interests or to resolve conflict.

In some ways the political debate between Anywaa and Nuer seems to be ironic. At a time when the entrenched, albeit narrowly defined, Ethiopian national identity has been deconstructed, it looks as if it is being revived at the periphery, albeit ideologically, either to justify exclusion from or as an argument for inclusion into the identity game. It is a double irony to hear the Anywaa and Nuer competing to 'be more Ethiopian' than the other, while both are 'excluded' from the national identity, based on colour discrimination.[32] Why should 'being' or 'becoming Ethiopian' matter for people who have always been at the fringes of a polity? For Anywaa, it is a way of preserving their identity and attaining a politically comfortable status in Gambela, which they consider to be their home. Southern Sudan is not only an uncertain and imagined polity, but is also perceived to be an entity to be dominated by the larger neighbours, the Dinka and Nuer. Such sectional interests are projected onto the national level in order to access and mobilize the state on their side in what is essentially a local struggle. Hence, the Nuer not only are a threat to the Anywaa but also to the Ethiopian state itself should they identify with an alternative southern Sudanese national identity. There is also a growing 'statist' approach among educated Anywaa, the belief that Anywaa society could only be rescued from an impending 'extinction' through the agency of the state. This is independent of the creeping influence and control of the central government/highlanders and a state that guarantees security from Nuer encroachments. Such an approach harbours an ethno-nationalist discourse, once championed by the GPLM but now vigorously advocated by an opposition party, the Gambela People's Congress.

For the Nuer, to be Ethiopian means unrestricted access to riverain lands and greener pastures at a time when population pressure on resources is becoming acute on the Sudanese side. To be Ethiopian could also mean to be dominant locally, via

---

32. Anywaa and Nuer face social discrimination when they travel outside Gambela, especially in Addis Ababa, where they are screamed at as *bariya* (slave) or euphemistically called as *lema*. The image of 'black' people in the capital is synonymous with southern Sudan; hence there is indiscriminate categorization as foreigners.

the politics of numbers in the regional identity game. Political reorientation seems to be related to disillusionment with seemingly endless conflict in southern Sudan, where a liberation ideology once raised hopes and allegiances from the various communities on both sides of the border.

Concluding Remarks

Despite the intensity of the conflict between Anywaa and Nuer in Gambela and their representation as 'traditional' enemies in the essentialist public discourse or conspiratorial theories of Nuer expansion, my findings reveal that whatever animosity exists and the various instances of 'ethnic' conflicts are expressions of a 'modern hate', the seeds of which could be traced in living memory. I disagree with the claim that ethnic conflicts have deep primordial (tribal, religious) roots. When contemporary Anywaa and the Nuer are in conflict, they fight over specific and concrete issues. The objects of the struggle are either access to vital resources, control over means of social advancement (particularly education), or attaining political power (capturing a regional state).

The intensity and the seriousness of the contemporary struggle, with an invocation of past memories, give the semblance of continuity from past conflicts to this 'modern hate'. The post-1991 political struggles between Anywaa and Nuer are fought against a historical background (an eastward pastoral migration not only militarily but also through an effective system of assimilation and social networking), differential evaluation of groups by the state in the regional power game, the impact of the civil war in southern Sudan, which created a new opportunity structure (determined the pattern of switching and reversal of groups' state identities), and the new post-1991 political structure in Ethiopia, which has enhanced ethnicity within a redefined nation state identity.

Proud of their seniority in Gambela and their competence in Ethiopian mainstream culture ('highland culture'), the Anywaa have legitimized their seizure of power, not only to dominate the present but also to redress old scores as well as determining the future in their own favour through a policy of containment. Such complex sets of interaction have generated a debate on who is who. The Anywaa use various ideologies of exclusion, initially as a defence for what they call the 'Nuer threat' but later on take the offensive. Autochthony, authenticity (genuine and reliable citizens) and land size are used as ideologies of entitlement and ownership rights over the Gambela regional state, while the Nuer contest the claim and, by implication, the charges against them with a longer historical frame and their own version of an origin myth (emphasis on the common origin of Anywaa and Nuer in southern Sudan). They also draw on the politics of numbers (manipulation of the census) and historical connections with and economic contribution and strategic importance to the national state. But perhaps the most important argument for inclusion in national identity is their contribution to the state in its war against invaders. Past contributions such as that during the Italian war in the Second World War are invoked, and more recently the current border conflict between Ethiopia and Eritrea has also been instrumentalized in the local politics of recognition, exemplifying an interesting fit between micro- and macro-politics.

Beneath ethnicity, what we observe in Gambela are multilayered concerns put in various interacting contexts. What is dubbed as a continuation of the traditional, age-old, natural enmity between Anywaa and Nuer contains within itself various interest groups. For the Anywaa and Nuer elites, the conflict is over who should control the newly constituted regional state. For Anywaa students, it is contesting differential access to quality education. For Nuer students, it is a matter of the constitutional right to learn in their own language. For the Nuer agro-pastoralists, including both spontaneous internal migrants and the recent influx from southern Sudan, it is a matter of access to natural (pasture, riverain land) and modern (goods, services) resources. For the Anywaa farmers, except in areas where inter-ethnic relations have been heavily politicized through a discourse of existential conflict and the resultant anxiety, it is an aspect of forming friendship networks to supplement a declining livelihood strategy and creating a buffer zone through social and military alliance with their immediate Nuer neighbours, which protects them from less familiar groups of Nuer who covet their riverain land. For the regional power holders (government of Sudan, government of Ethiopia and southern Sudanese rebel movements), it is a strategic concern to mobilize one group or the other, depending on a relative evaluation of group worth, to fight their own war.

As I have shown, the major conflicts throughout the 1990s in the Gambela region are in the political domain. The state is used as an arena within which the local struggles are fought. The paternalist, non-transparent federal state has further compounded the situation, in which the local elites manipulate group identity in their competition for leadership. A more focused analysis and observation reveals that what is at stake is not identity per se but using different strategies of entitlement and means to exclude others from political power and vital natural resources. The masses are invited to activate their ethnic identity and join the struggle with the pretence that they have a stake in it. The contemporary struggle is framed in emotional terms: selective memory from past conflicts, particularly politicization of the age-old pastoral expansion in the region. This has made the request for political mobilization appealing to the ordinary people. As a result, a growing number of the masses are developing a stake in this new ethnic identity game. Education and the language question have become the missing links between the elites and the masses. The political setting has created an incentive for education as a means to having access to new resources such as leadership and employment opportunities. In fact, the state itself is one of the objects of the struggle. The ethnic groups now use different strategies to access and seize the state in order to determine the outcome of the struggles in their own favour. The net result is an emerging social closure in a region where fluidity has been the norm.

Chapter 9

The Cultural Resilience in Nuer Conversion and a 'Capitalist Missionary'

Christiane Falge

Introduction

This chapter deals with the development of Nuer Christianity in Ethiopia during different phases of conversion, from the arrival of the first missionaries in the 1960s to the establishment of a Nuer branch of the EECMY (Ethiopian Evangelical Church Mekane Yesus) in the 1990s. It discusses different aspects and effects of conversion, and explores the impact of encounters with missionaries on Nuer society. Rather than assuming a single, central motive for conversion or a particular, specific effect of the missionary encounter, I draw attention to the cultural resilience of the Nuer in their conversion, on the one hand, and stress the variety of forms that Christian conversion can take in different contexts, on the other. I argue that the missionary encounter was only one of many different cultural conversations that have taken place and is only one kind of global influence that has shaped the Nuer in their encounters with various world narratives; it should not, therefore, be overemphasized. I demonstrate this by describing the different reasons for conversion and by showing how the meaning of Christianity has changed for the Nuer during various different phases of conversion.

Approaches to Conversion

I start my chapter by referring to three approaches to understanding conversion. The Weberian approach is among the most common explanatory tools and one to which many later theories of conversion and social change refer. It assumes that the Protestant ethic provides the universal path towards modernization. In their historiography of Tswana conversion, the Comaroffs argue that the long conversation that took place with the missionaries changed 'Tswana minds', in the sense that it generated an unconscious surrender to capitalist hegemony. The Comaroffs differ from the Weberian approach in that, while Weber assumes that dogma and revelation through the puritanical work ethic will adapt people to a capitalist hegemony, the Comaroffs explain it independently of dogma and put more emphasis on the long conversation between missionaries and converts: 'While the missionaries did not succeed in converting many Tswana, they were successful in

implanting in them the cultural forms of bourgeois Europe on African soil'
(Comaroff and Comaroff 1991: 311).

Donald Donham criticizes this approach to missionary encounters for
homogenizing missionary and Tswana 'cultures' and for attributing agency to
abstractions rather than to people acting in particular material contexts (Donham
2001: 134). In his own detailed ethnography of the Maale of southern Ethiopia, he
rejects the idea of local peoples surrendering to capitalism through a long
conversation, and explains it instead as a result of the particular position of a country
within a global system dominated by capitalism (Donham 1999: 247). He describes
how the Maale actively converted to Protestantism because it was a way for them to
escape the marginalization and stigmatization they experienced. Previously they had
been seen as 'primitives' and non-orthodox Christians and not as 'true' or 'full'
Ethiopians. Although the Comaroffs had no intention of essentializing Tswana
culture, they nevertheless overestimated the missionaries' influence.[1]

By giving some insights into Nuer conversion, I show here the relative
insignificance of the encounter with missionaries compared with the development of
a 'capitalist spirit' among the Nuer, and draw attention to the agency of the Nuer in
their decision to convert. The missionaries failed to convert many Nuer, and most of
their converts were as little interested in dogma and revelation as the Tswana or the
Maale. Rather than simply accepting and adapting certain 'cultural forms of Europe',
conversion had multiple effects and had various meanings, from resistance against
power to hope for a better life through modern technology or spiritual healing.[2]
Christianity also provided the Nuer with an alternative identity that compensated for
the loss of meaning experienced during the civil war in Sudan. The Christian view
rejected Nuer violence and military ethos, as well as officially rejecting the Nuer
divinities and prophets[3] and stigmatized cattle thefts and killers. In the following I
look at these different dimensions of Nuer conversion, and assess the way in which,
over different periods of time, they resulted from encounters with missionaries or
from other 'cultural conversations'.

I explore the first two phases of Nuer conversion. The first Nuer conversion
happened in the context of an early spirit of resistance to the Islamic government of
Sudan. It was characterized partly by a discourse against Arab oppression and partly
by a fascination with Western knowledge, superior technologies of healing and peace.
The second conversion took place in the 1980s, after war broke out again following
a decade of peace. Here, Christianity increasingly became an alternative form of
identity in the war-affected society. The Ethiopian Nuer were less directly affected by

---

1. John Comaroff hinted at this overestimation during a personal communication with the
   author.
2. For similar motives for conversion as a way to 'improve one's life' see Ren'ya Sato among the
   Majangir (2002: 187) or Kaplan (2005: 112).
3. Nuer prophets were once introduced to us by Evans-Pritchard and others as a form of anti-
   colonial resistance and as protectors of Nuer territorial boundaries against neighbouring tribes.
   They became increasingly specialized as healers when military protection was taken over by the
   SPLA.

the Sudanese war, but they experienced its effects in refugee flows and SPLA military camps. For them, conversion became a means to 'becoming modern Ethiopians'. This form of conversion became increasingly pronounced as larger groups of Nuer began to have access to modern technology and schools through the newly established refugee camps.

Phase One: Conversion and Resistance

Until 1960, missions in Nuerland were only established on the Sudanese side and very few Nuer had converted to Christianity. The war forced Sudanese Nuer to take refuge in Ethiopia and, in the early 1960s, a semi-formal refugee camp was established in Pilual.[4] In 1964 the refugees moved to Odier, where they joined the rebels' settlement. Pilual and Odier are highly significant places in Nuer narratives of conversion and they came to be known as the first 'Christian places'. Many evangelists who had fled from Khartoum, Nasir and Akobo settled there together with Anyanya-I soldiers: 'The movement [Anyanya-I] was already in Pilual when the evangelists came. For them it was an opportunity to set up the word of God among the people who were in the wilderness[5] or in the camp' (Duoth Dul, Addis Ababa, 2004).

Although the leaders of the Anyanya-I movement were mission-educated Christians, in the early days of the movement most of their soldiers were non-Christians. When the Anyanya-I soldiers settled in Pilual, they lived together with Nuer Christians who had established the first Presbyterian Church of Sudan (PCOS) in Ethiopian Nuerland. This proximity led to the further conversion of many soldiers to Christianity: 'Initially their aim was to fight for the freedom, not for the gospel. They got the gospel in the camp as an opportunity for them. But before, they didn't know anything of God' (Guarner Chuol, 2001, Iowa).

Many Nuer converts told me about their encounters with Anyanya-I soldiers and that one of the main factors in their conversion was the association of the liberation movement with Christianity. Many of them encountered Anyanya-I soldiers marching past singing hymns like 'The Bible is our gun', and were inspired to convert. As many evangelists became soldiers and many soldiers converted to Christianity, soldiers, refugees and Christians came together under an overarching Christian identity with the aim of countering their marginalization. 'During Anyanya-I we were all Christians and we would always put God in front of fighting. Soldiers used to go to church. We prayed before fighting. Even our wives were deacons and we had evangelists among ourselves. Christianity was part of the struggle against Muslims. We constructed two churches in Bilpam and our leader was pastor Ret Lony' (Stephen Kuey Mayan, former Anyanya soldier, Fangak 2004).

Up until the 1970s, therefore, being a soldier was equated with being a Christian. Conversion to Christianity was seen as a political decision that simultaneously converted a person into a rebel. In the words of Reverend Duoth Dul:

---

4. The refugees, who were predominantly Nuer, were provided with grain delivered by a steamer known as 'Baro Nesh', which sank in 2001.
5. 'Wilderness' (*duor*) refers to any place outside Nuerland where people are forced to go when in a state of anomie like war.

Those who came from the districts that were destroyed by Arabs were all together in Itang. Then they continued to pray to God – blaming the word of God. Some people who didn't hear the word of God learned from the Christians and accepted Christianity. And then the most important factor, accepting Christianity, was the fact that 'coming out' [coming out refers to becoming a rebel] 'to be a refugee' meant to join a religious movement. (Rev. Duoth Dul, Addis Ababa, 2004)

The association of conversion with resistance to the 'Arab colonizers' and the all-embracing Christian identity of soldiers and laymen changed during the second phase of conversion with the increasing escalation of the war.

Conversion and healing

A further reason for conversion during the initial phase was the emergence of various healing practices. The practice of Christian healing was introduced to the Nuer on the Sudanese side in the 1960s by the American missionaries in Nasir and on the Ethiopian side in the 1970s by Nyatap, a Nuer deacon who had worked with the missionaries in Nasir. Christian healing practices were an act of resistance against and induced a shift of power away from the prophets to the community, as they empowered people to heal themselves. Prophets related sickness to a deteriorated relationship with God or members of the community, and they healed people through cattle sacrifice. It was a method that required the patient to contribute sacrificial cattle in order to appease God and lift the affliction. Fascinated by the fact that 'Jesus could not be bribed', as Nuer converts tended to say, many converts enthusiastically began to practise Christian healing 'free of charge'. Unlike prophetic healing, which needed the prophet/spirit possessor to act as a mediator between the divinity and the patient, Christian healing was based on direct communication between the patient and God. It attributed agency to the patient or the healing group. Christian healing sessions were often practised under the guidance of a church leader, frequently a female deacon.[6] The healing group aimed to accumulate divine power through prayer and to channel it into the sick person. Nyatap, who became a famous healer, explained to me how she developed her healing skill:

Healing came before Manpiny[7] from the first people who brought Christianity. It is called 'fellowship'. Some people are still alive today, because of that praying. We were nine people when we were still in Nasir. There was a small boy. He was taken to hospital, but the doctor told us to

6. One of the reasons why Christianity experiences a high number of female converts is the important roles Christianity offers to them as healers and church leaders – roles that they assume only rarely as non-Christians.
7. Manpiny was the Nuer missionary Charles Jordan on the Ethiopian side, who is discussed in more detail later in this chapter. The name was given to him by the Nuer and can be translated as 'the one who covered it all'.

take him back, that he was dead. We sent some people to call others to bury him. Then we surrounded the boy, the nine of us, the ninth was our pastor. One after the other prayed, until it was the pastor's turn and the boy started crying. He is still alive now. When we came here, a woman was sick. In the evening she became unconscious. People started crying, because she was dead. We took her to the other side of the river to bury her. Others said, let us just pray and keep her in the house. Then we carried her into the house. In the morning we woke up early to go to her house and found her sitting with her brother. I realized that it was the [work of] God. And, if it had been a prophet, he would have taken the cow. Now we have known the difference. God is good. (Nyatap, Adura, 2002)

These healing practices still form a substantive part of Nuer Christianity today. Nuer Christians practise them alongside Western medicine.[8] They constituted an important aspect of conversion, even when missionaries began to introduce Western medicine. A former prophet who had converted to Christianity in the 1970s explained how his conversion miraculously enabled his wife to recover from infertility: 'When I was a prophet I lived with my wife for a long time without giving her a child. They said that the reason was my God. When I came here [Adura mission] and became a Christian, my wife got pregnant and delivered a child' (Joschua Chen Kruth, Adura, 2002).

In the face of the high level of HIV/AIDS infection rates among the Nuer in Gambela and the simultaneous absence of antiretroviral drugs, all types of healing are currently gaining value. Recently, the mother of a friend who is an AIDS patient was called to his assumed deathbed in Addis Ababa. When the mother, who is also a deacon, arrived at her son's house, she took off her deacon's uniform and exchanged it with her son's pillow. The son revived miraculously[9] and Christianity once again manifested its power.

Several authors mention the parallel existence or slight modification of customs based on local belief systems by new religions. Kaplan, for instance, argued that, rather than replacing them, Christian ideas and practices were added onto those of old faiths (2005: 111). Similarly, Nuer healing is not completely detached from 'cattle sacrifices'. Cured patients often thank God by donating a cow, *yang kuoth* (cow of God), to the church. This is seen as distinct from non-Christian sacrificial slaughtering. A cow that is cut with a knife is considered to have been slaughtered and to be suitable for Christians to eat. A cow that is killed with a spear is considered to have been sacrificed and the meat is only suitable for non-Christians.

---

8. It should be noted, however, that this type of healing does not have the central significance to Nuer Christianity that it has in many African Pentecostal communities. Rather, healing is just one among many important aspects of Nuer Christianity.

9. My friend had, however, also gained access to antiretroviral drugs shortly before his mother had come.

The establishment of the Nuer mission in Adura

In the same year that the Pilual refugee camp was established (1964), Charles Jordan, a missionary from the Presbyterian Church of America arrived in Ethiopia. Known locally as Manpiny, he established the first Nuer mission in Adura, which became a place of refuge for many Nuer refugees.

> The time the missionaries came, Adura was the first place to be established for the missionary station. The founder of the Nuer evangelization was Jordon Manpiny. He came as a layman at that time and he taught the first converts. As they accepted Jesus Christ as a personal saviour he called Reverend Reimer, the Anywaa missionary who baptised Manpiny's first converts.[10] (Rev. Duoth Duol, Addis Ababa, 2004)

The arrival of the missionary was accompanied by the opening of the first health clinic, which led to further conversions among the Ethiopian Nuer. The benefits of medical work had been praised enthusiastically by Dr Thomas Lambie, one of the pioneers of the Sudan Interior Mission, who considered it to be the most efficient tool for conversion (Donham 1999: 88–89).

Charles Jordan's patients were taught Christian prayers while undergoing medical treatment. After the injured[11] or sick were returned from the hospitals further afield, they would stay in the mission clinic until they made a full recovery. Many converted at this time, after experiencing church services and attending Bible school training while at the clinic (Gatdet 2000: 30). In 1973 and 1974, Jordan ordained the first four Ethiopian Nuer pastors and some elders[12] who then established the first congregations in Mangok, Jikow and Pilual.

The Nuer did not have a strong presence in towns at this time, and the Anywaa, who had already begun to convert and become literate, began to derogate and exclude them as 'backward Sudanese refugees', referring to their lack of formal education. Conversion to Christianity provided a way for the Nuer to escape this image, as converts were also enrolled in the missionaries' literacy classes. The missionaries had a preference for living in towns,[13] a preference that was related to ideas of salvation. The hardship in rural areas was perceived as an expression of Satan's victories. Missionaries, by spreading the word of God in those areas, would

---

10. Jordan was ordained as a pastor very late and converts were usually baptized by Nils Reimer, the Anywaa missionary. The fact that Jordan was ordained only after he took up missionary activities contributed to the Nuer suspicion about his 'real' intentions and reasons for being in Nuer areas.

11. In many cases, men with injuries caused during sectional fighting were transported to Adura clinic and from there were flown to clinics in the highlands.

12. These were all people who had converted in Adura mission.

13. Unfortunately, neither the Reimers nor the Jordans have published accounts of their missionary activities. I did, however, meet the Reimers in Gambela and communicated with them and Jordan by email, and they explained to me about their missionary activities in Gambela.

ultimately bring the end of time nearer. Most evangelical missionaries outside the central urban core felt that by retreating to remoter rural areas they could fight against Satan better; the 'domestication' of the 'remote', 'local' and 'wild' was perceived as a victory over him (Comaroff and Comaroff 1991: 68; Donham 1999: 10). This missionary association of rural, non-Christian, uneducated people with the 'primitive' served to reinforce Nuer discourses about the 'primitive' as uneducated and rural, and the 'civilized' as educated and urban, a set of associations that had already been introduced in colonial times.

Conversion and the source of 'modern' knowledge

The missionaries impressed the Nuer and Anywaa with their machines, such as aeroplanes, grain mills and fridges. These encounters with new technology confounded the Nuer about what was animate and inanimate and what was man-made and what God-made, and, in trying to make sense of it, they linked the missionaries' commodities and machines directly to God. This link was based on a belief that technology resulted from a mysterious tie between the Christian God and a praying human that would enable divine knowledge and wisdom to be inculcated in the praying individual. As for the Nuer, it was this flow that explained the mysterious commodities and power of Western knowledge.[14] Praying and hymn-singing, therefore, became a performative act through which they imagined they would accumulate this knowledge. Western medicine and modern assets were perceived to be part of the missionaries' God, a power with which the Nuer prophets were unable to compete. Its success influenced many people's decision to convert to the 'superior religion'. Interestingly enough, the Nuer believe that the knowledge of technology originates from them, but had been stolen by the British colonialists. Conversion to Christianity, as some Nuer Christians believe, enables them to accumulate flows of this knowledge by establishing divine links with the Christian God through prayers.

All missionaries and nurses that worked in Gambela used aeroplanes for their work in Anywaa and Nuer areas, the latter of which are flooded for about six months of the year. Many church-related events were associated with technology and the fact that the missionaries were moving with planes as if they were the 'whites' common vehicles strengthened the belief in the divine source of that knowledge. It was not the technology itself that was believed to be divine but the knowledge, which required human agency linked to God in order to generate these commodities. In the Nuer model, humans were attributed with agency and the ability to implement God's ideas. This is different from the cargo cults, where agency rests solely with God (Lawrence 1967). These ideas about technology are in line with Nuer 'egalitarian' ideas, because the only thing that is needed for access to knowledge and hence the technology was belief – something that is accessible to everybody.

The explanation of the divine origin of Western knowledge turned the anti-modernist Christianity of the missionaries into a modernist movement, because

---

14. This idea of the knowledge tie between God and a Christian resembles the Nuer idea of a mysterious blood tie created between slayer and slain that would pollute the slayer unless he cleanse him- or herself of the homicide.

Christianity became the transmitter of modern commodities. Donham described similar incidents among the Maale during the 1920s when the premillennial SIM (Sudan Interior Mission) focused on impending crisis and ultimate salvation, believing that high conversion rates would hasten the second coming of Christ. The founder of the mission welcomed modern technological advances and hoped to use them to accomplish his anti-modernist religious end. The deliverance of modern facilities like schools and hospitals became one of the main push factors for Maale conversion (Donham 1999: 85–95).

Until the 1970s, schools and churches were only found in towns and missions so that Christianity was associated with education and urbanism. The Christian God was also addressed as *kuoth turuk* – the God of the knowledgeable.[15] The idea of knowledge transfer through prayer was slowly replaced by the idea of literacy as a skill needed for an urban life, and a skill that would open the path to a new world, to which an increasing number of young Christians began to aspire.

Further cultural conversations with world narratives

Despite the obvious emergence of an alternative Nuer identity among converts, this identity was constructed by themselves, and, as we have seen, it was, at least in part, different from that intended by the missionaries. Considering the geographical distance between their settlements and those of the missionaries, as well as the fact that the encounter was limited to a few hours per day, it is difficult – in the Nuer case – to imagine that missionary narratives were central to these innovations. Rather, the Nuer appropriated some of the missionaries' elements in different ways and for different purposes. Among the Nuer, one of the most valuable attributes of a leader is social interaction and proximity to people. The fact that Charles Jordan seemed to be uninterested in establishing bonds of friendship gave him an aura of secrecy. In local discourse, Jordan was perceived more as an economic entrepreneur than as a successful leader. Many Nuer stress that he was not a pastor, but more like a trader who was guided by individual economic interest. Some recall that he bought grain for low prices during lean periods and sold it again to the Nuer for a high price when their stores had been used up. Jordan's other sources of income included a hydraulic grain mill, a sewing workshop, fishing boats and several cattle byres for cows that he had acquired through his involvement in the cattle trade or from people who exchanged cattle for grain.

Furthermore, the growth of Jordan's economic wealth and, most of all, the magic of his seemingly 'inexhaustible' grain store simultaneously attributed power to and aroused suspicion of him. The Nuer felt that, for some reasons, Jordan excluded them from the source of the knowledge that constituted the key to his tremendous wealth. This suspicion partly emerged from their colonial experience, which made them perceive him as an extractor of their power, which he transformed into 'magic

---

15. *Turuk* or in its translation 'the knowledgeable' can also refer to outsiders or white people. Nowadays it is used for anybody with a 'modern' knowledge. I do not refer here to 'modernity' in the Western sense, but to 'modern' knowledge, power and technology, such as that related to aeroplanes. Literally, of course, *Turuk* means 'Turk'.

wealth'. This vernacular explanation of the relations of domination, emanating from the nature of the unequal exchanges between the missionary and the Nuer, was explained to me by a former convert who accused him of prioritizing his own interests over the development of Nuer society: 'He did not develop the area ... The grain store was his secret. He exchanged grain for cattle when people were out of grain ... In his shop people would exchange grain for commodities. Manpiny had several big cattle byres and many fishing boats' (Bol Kiir, Des Moines, 2003).

Charles Jordan's presence in Gambela is, therefore, remembered foremost for his capitalist spirit. Nuer narratives recounted the continuous growth of his grain stores and the number of cattle byres and fishing boats. Rather than explaining his accumulation of wealth rationally, however, or internalizing it by adopting similar forms of accumulation, they related these activities to a Christian God. The Nuer were also attracted by the missionaries' technology, knowledge and power, because they saw the improvement this technology would bring to their lives. This aspect of conversion can be seen to some extent as an adaptation to a certain capitalist ethos. Today this ethos expresses itself in material interest in certain Western goods that the young Nuer especially demand from the diaspora or purchase from urban areas. Importantly, however, 'money-making as an end in life' as one of the criteria of a spirit of capitalism (Weber 2001) has not been adopted by the majority of Nuer. Only a few Nuer have managed to escape the local exchange spheres and participate in an urban market economy. Today only one of the approximately thirty restaurants in Gambela is owned by a Nuer, and very few Nuer in Ethiopia are involved in entrepreneurial activities. Although they sell their cattle and each week large herds of cattle are being taken from Jikow to the cattle market in Gambela town, these cattle are sold for the sake of immediate needs, like medicine, education or clothes, but not with the sole aim of accumulating money. The importance of cattle for the Nuer political organization, marriage and social security and, most importantly, for the strong claims of kin does not allow them to become fully involved in an urban economy, and the majority of their wealth continues to lie in their herds. Even salaried, urban Nuer face great difficulties in saving money or becoming involved in trade owing to the reciprocal claims that relatives from the rural areas continuously make. There are also other reasons, related to the power and position of Ethiopia in the world market, that do not allow a capitalist spirit to develop among the Nuer, which we cannot, however, delve into here.

In 1972, new events shaped the history of Nuer Christianity in Ethiopia. The Addis Ababa Agreement was signed, and all Sudanese refugees and rebels returned to Sudan. By the end of the first conversion phase, missionary encounters had neither led to a considerable transformation of Nuer economy nor aroused the interest of a large number of Nuer to convert. They had, however, triggered an interest in modern technology and wealth, as well as evoking the idea that accessing this knowledge was dependent on conversion and literacy.

Conversion after the Missionaries

The heterogeneity of local discourses

After the socialist revolution in 1974, the Ethiopian regime expelled all US-based institutions, including the missionaries. Charles Jordan and his wife had to pack and leave the same day. This event and its aftermath are remembered in several contradictory narratives. The reason for their presentation here is not to find the 'truth' but to get away from a homogeneous representation of people's perceptions of various encounters and the effects they have on people. The various memories of the missionaries' departure reflect the views of different groups of Nuer after their missionary encounter: lay Christians and church leaders. A repeatedly told narrative from Nuer living around Adura mission, for example, is about Charles Jordan burning his property on the mission compound, instead of handing it over to the Nuer. This memory reflects the attitude discussed above, that many Nuer saw the missionaries as people who denied them power and wealth.[16]

In another narrative, Christian laymen blame Jordan for handing the mission over to the government, leaving Nuer Christians without a base. In a third narrative, however, Nuer church leaders stated to me that the Nuer lacked loyalty to the missionaries, and that this was the cause of the loss: 'The missionaries left for their own land. That happened because the Derg was communist and controlled the Ethiopian government. These people didn't get any room. Of course, for those who defended their missionaries, they were allowed to stay, but for us [Nuer] nobody supported them' (Rev. Duoth Dul, Addis Ababa, 2004). Rev. Duoth Dul's narrative continues with the visit of a government delegation to Gambela after the missionaries' dismissal, in which the Nuer were asked whether they wanted the mission to remain with the Ethiopian church Mekane Yesus or be nationalized. By voting for nationalization the Nuer Christians lost the mission – a narrative that explains the loss with reference to their disloyalty and defamation of the missionaries.

Rather than a reflection of a single encounter with the missionaries, the various memories of the departure of the missionaries reflect the encounter that the Nuer had with various world narratives. Each narrative contains memories of conquest, trade and exchange in the past, as much as it relates to the narrator's structural position within society. The laymen's memory blames the missionaries for the Nuer lagging behind and reflects their continued experience of marginalization. The church leaders' version, which blames the Nuer themselves, has a clear Presbyterian 'touch' and has been shaped by their missionary encounters. The church leaders' narrative also creates church loyalty and serves their interests as the current 'missionaries' by referring to disloyalties and the sins of the past. All three narratives share an acknowledgement of Nuer 'backwardness' or weakness and the power of the church or Christianity. Out of this has emerged the pressing need to 'catch up' with those with superior knowledge and power.[17]

The Nuer narratives of the departure of the missionaries contain elements that are reflected in other narratives from colonial times that thematized Nuer

---

16. In an email to me Jordan himself denies that he burnt his property.
17. The 1970s were a time when the state could hardly provide services in the centre, and the church was the only institution that delivered services to the periphery. The state was therefore an object of major contestation. The church, as a non-local institution, was perceived of as having government-like authority, and competed with the state and state-like institutions. This situation largely remains the case today.

'backwardness' as a cause for their surrender. The motive of self-infliction, on the one hand, and the motive of external infliction due to outsiders' extraction or denial, on the other, are interchangeably applied to explain the experience of loss and domination during colonial and current times. Christians usually end these narratives by concluding that conversion is the only means to escape this perceived 'backwardness'.

The active engagement in conversion

Emancipation from the demanding earth divinities that possessed the prophets was another important aspect for conversion in the late 1970s, as explained to me by Nyatap, a famous evangelist from Adura mission. To Nyatap, Christianity caused the demise of prophets, whom she views as mere exploiters:

> The prophets are desperate. Nobody listens to them. It is the end of everything [*Ca doar ngok*]. Things are good when they are going ahead [*Ba gwa ka wa nhiam*]. People were tired of these 'Satans'. If you did not bless this God to the evil spirit some one will die in your family and then people will say, it is because you did not give a cow to God. Then we heard that it is one God whom people are worshipping. When people tried Christianity and others saw that they didn't die, they are still alive, they trusted the real God. They said that Christian life is good, because there is nothing you have to offer to God. All people opened their hearts to God in heaven, because there is no one who will ask for your life. (Nyatap, Adura, 2002)

Nyatap's view of conversion and Christianity reflects her close attachment to the Jordans as she was one of the few Nuer who lived in their compound. The majority of cases I encountered disproved this view of a radical break with the past. Most converts remain connected to the past by practising both belief systems, which is in line with Kaplan's argument that conversion to Christianity often adds to local belief systems but does not replace them (Kaplan 2005: 111). I frequently met Christians consulting prophets in search of solutions to diseases or social problems. These Christians appeared to have no problem with the simultaneous adherence to both belief systems. The prophets were being consulted a lot as they have a wider spectrum of intervention than Christians. While in the early euphoria of the 1970s Christians were more radically turning against prophets, today they are more tolerated as people combine several belief systems for their various physical, psychological and social problems.

In most cases, Nuer conversion is not about a break with the past, in a sense that earth divinities are dismissed as unreal. Rather than ceasing to exist, they may lose their power against Christians but continue to affect powerfully those who believe in them. However, one can never be sure of not being affected by them. A prophet who converted to Christianity had actively to free himself from his divinities. The former prophet Joshua Chen Kruth explained his experience as follows: 'After I became a Christian the divinities kept coming. They came at night and when I woke up I prayed. They came and tested me again. Later, I would pray again. Always like that, until they disappeared' (Joshua Chen Kruth, Adura, 2002). But, even in this case of a total rejection of the past, the divinity only loses its power over the individual. It does not cease to exist. Among the majority of Nuer converts, the two realms, the Christian and the non-Christian, coexist.

Cultural resilience and the transformation of the Nuer church

The idea of a Christian community as introduced by the missionaries aimed to transcend Nuer kinship and create a new Christian family, detached from local culture. Hence, during the 1970s when this idea was prevalent, many Christian settlements in Ethiopia were formed by Nuer refugees from all corners of Nuerland. The most famous one is the village Jerusalem near the town Jikaw, formed in 1972 by refugees returning to Sudan. These people originated from various Nuer clans and, rather than returning to their home communities, decided to live together as a Christian community. This idea of an overarching Christian identity that transcended lineage identities was combined with ideas about equal access to knowledge and power through divine links. The unifying character of a Christian identity was, however, short-lived and, along with increasing conversion rates and the re-emergence of war in the 1980s, this idea ceased. Nuer lineages became salient again. In line with the re-emergence of lineage over Christian identities, ideas about divine flows of knowledge were added to by ideas about the necessity of access to schools and the passing of examinations. Furthermore, Nuer Christianity began to take up increasingly elements of local customs. The solemn, dark, slow missionary hymns were replaced by vibrant Nuer compositions that were faster in rhythm and brighter, as well as louder in tune. New churches sprang up with new hymns and dancing styles that increasingly embodied local dancing practices. The Nuer military ethos also emerged: the Bible was described metaphorically as a 'big gun' (*mac in diit*) that Christians carry to fight against the 'beholders of evil' (*gwen jekni*). Christians referred to themselves as 'a fighting force' (*ran köör*).

Soon, the number of pastors was no longer sufficient for the expansion of Nuer Christianity, and lineages started to compete with each other over the number of converts and for pastors as signifiers of their power. The church offered access to relief, scholarships and government jobs.[18] In this context, lineages and what I describe elsewhere as segmentary Christianity[19] (Falge 2005) became the most important tools for the spread of Nuer Christianity. This importance of the segmentary lineage in a new context challenges some of the anti-structural functionalist critiques of Evans-Pritchard of the 1980s.

In the newly established Ethiopian and Kenyan refugee camps, which became the centres of emerging Nuer Christianity, Nuer refugees established an administrative structure that represented the PCOS (Presbyterian Church of Sudan) in the camps.

---

18. The relation that exists between government jobs and conversion is based on the way the churches were the main entry points to education as they were the only institutions in remote Nuerland that offered literacy classes.

19. Sharon Hutchinson once mentioned this term to me during a conversation on Nuer Christianity.

In 1989 they formed the Nairobi-based New Sudan Council of Churches (NSCC) as a sister organization to the Council of Churches in Sudan (SCC), as the latter operated solely in areas controlled by the government of Sudan. As an umbrella organization for all churches, the NSCC is coordinating development-oriented church activities in the whole of southern Sudan. By becoming one of the most powerful southern Sudanese institutions sponsored by the West, it linked up southern Sudan with the Western world. Christianity flourished in the camps and from there spread to other exile settlements and Sudan.

Conclusion

This chapter started out by looking at approaches to conversion that linked missionary encounters with the possibility of converts' adaptation of certain European cultural forms, such as the spirit of capitalism. This transformation was assumed to occur through the power of dogma and revelation in the Weberian approach and through long conversations with the missionaries in the Comaroffian approach. I argue that the missionary encounter should not be overemphasized in reflections about social change but should be seen as one of many other cultural conversations and influences. In the era of globalization, religious movements increasingly adopt heterogeneous elements from many different places, so that the categories of local, global, indigenous and foreign hardly make sense any more. The comparison of two conversion phases shows the various forms Nuer Christianity has taken at different points in time, during the presence and the absence of missionaries. While partly adhering to a capitalist ethos by admiring Western forms of technology and power, the Nuer have neither fully monetized their exchange spheres nor made money-making an end in their lives. Christianity has also flourished in diverse ways since the missionaries left. Hence, rather than adapting to 'foreign' ways, this chapter concludes that the Nuer have appropriated Christianity in their own ways.

Chapter 10

Changing Identifications among the Pari Refugees in Kakuma

Eisei Kurimoto

Wars in North-East Africa have produced millions of refugees and internally displaced persons – many more than the official UNHCR figures reveal. They live under a variety of conditions, in refugee camps and settlements under the protection of the UNHCR, in camps for the displaced often without the UNHCR protection and in shanty towns on the outskirts of urban centres in utter destitution. Although there is an enormous volume of reports and documentation on refugees, we know very little about their livelihoods and social worlds. In other words, most of the reports and studies recognize them as abject, helpless and hopeless victims to be rescued, resettled and repatriated. Only a few treat them as individuals trying to survive and to reconstruct or construct a social world in which to live.[1]

This chapter is a preliminary account of Pari refugees who were involved in the Sudanese civil war, became displaced and now seek refuge in the Kakuma refugee camp in north-western Kenya.[2] As I associated with them in their home villages and in Torit and Juba towns in south-eastern Sudan when I conducted fieldwork intermittently between 1978 and 1986, visiting them in Kakuma was primarily a 'reunion' after many years, both for myself and for them. Here I try to account for how they have survived the turbulence and how, in this new setting, they manage to maintain social ties and to reconstruct a community. I argue that, regardless of the seemingly 'cosmopolitan' nature of the camp, there is a growing consciousness of being Pari and the momentum of their identification is inward-looking, with persistent references being made to the 'Pari community'.

---

1. Liisa Malkki's (1995a, b) work is a major exception.
2. Research conducted in Kakuma during 20–27 February and 9–23 September 2000. The first research, which I conducted with Dr Peter Adwok Nyaba, was sponsored by the Suntory Cultural Foundation, and the second by a grant-in-aid from Monbusho (the Ministry of Education, Science, Sports and Technology, JaPari). An early version of this chapter, 'The Multicultural and Multinational Kakuma Refugee Camp: a Space for a Hundred Flowers to Bloom?', was presented at the 'Culture and Development Workshop' held at Kyoto Seika University, 7 October 2000, on the occasion of Ngugi wa Thiongo's visit to Kyoto.

Kakuma Refugee Camp: Population and Basic Services

Kakuma refugee camp in the north-western corner of Kenya was established in 1992, primarily for Sudanese refugees, many of whom were unaccompanied minors. Until May 1991 they had stayed at refugee camps in western Ethiopia – Itang, Pinyudo (Fugnido) and Dima – but they were forced to run away from there because of the fall of the socialist Derg regime of Ethiopia and the coming to power of the Ethiopian People's Revolutionary Democratic Front (EPRDF). Many of them sought refuge in Pochalla, and others in Nasir in Sudan. In Sudanese territory they became the target of an offensive of the Sudanese army, and after a long trek they finally sought refuge in Kakuma. However, in the course of time the camp began to accommodate refugees from Uganda, Congo-Kinshasa, Rwanda, Burundi, Somalia, Ethiopia, Eritrea and so on, primarily because of the Kenyan government policy to relocate refugees from urban centres and from other camps there, making it a multinational refugee camp.

The refugee population in Kakuma has been steadily increasing, although there have been significant drops after 'headcounts' by the UNHCR sub-office. The population in December 1994 was 48,462, and the result of the 16 December headcount showed 27,825,[3] a dramatic decrease of 43 per cent. In February 2000, it increased to 87,000 and the result of the May headcount was 65,000, a figure decreased by a quarter. The official figures as of 13 September 2000 were 66,400, and their breakdown by nationalities is as shown in Table 10.1.[4]

Table 10.1 Kakuma refugee camp: population breakdown by nationality as of 13 September 2000

Sudanese	51,953
Somalis	11,501
Ethiopians	1,995
Ugandans	331
Congolese	203
Rwandans	177
Burundians	107
Eritreans	27
'Stateless'[1]	108
Total	66,400

[1] 'Stateless' refugees are those Somalis who were once repatriated to Somalia and then came back claiming that they are Kenyan, not Somali, nationals.

These 'drop downs' after the headcounts suggest a high mobility of refugees as well as the economic/political significance of population counts. As the amount of food each household receives is determined by the 'official' number of people who are indicated on 'ration cards', they try to maximize their number for their survival. Therefore 'counting refugees' is always a sensitive issue.

3. UNHCR (1994).
4. UNHCR (2000).

The Sudanese comprise an overall majority (78 per cent) of the total population, followed by the Somalis (17 per cent), and the Ethiopians (3 per cent). The top three share 98 per cent of the total, while the rest remaining five nationalities and 'stateless' share only 2 per cent. Nonetheless, the multinational composition of refugees is a remarkable characteristic of Kakuma. Moreover, the Sudanese, in fact overwhelmingly the south Sudanese, have a highly multi-ethnic composition, although the Dinka refugees form the majority. As of the end of 1995, the total number of Sudanese was 39,160, consisting of thirty-nine ethnic groups. The Dinka refugees made up 77 per cent (30,216) of the total south Sudanese population, followed by the Didinga (7.2 per cent, 2,829) and Nuer (6.8 per cent, 2,670), and the other thirty-six ethnic groups composed 8.8 per cent (3,445).[5] The ratio of the non-Dinka to the Dinka population seems to have been increasing with new influxes of refugees from Sudan.

It is undoubtedly an enormous task to maintain this camp as it is located in a very remote area. UNHCR takes overall responsibility and cooperates with implementing agencies (non-governmental organizations, or NGOs). The World Food Programme (WFP) is in charge of transporting food to Kakuma, at the rate of 1,300 metric tons every month. The leading implementing agency is the Lutheran World Federation/Department for World Service (LWF/DWS). It is responsible for the distribution of food, soap and firewood, the provision of drinking water and materials such as sun dried bricks, poles and sheets to construct shelters, the construction of latrines and infrastructure, the implementation of education programmes and the provision of various community programmes such as psychosocial care, gender, conflict resolution, capacity building and income generating and sports, youth and culture. Its total expenditure in Kakuma in 1999 was about $4.2 million.[6] Other implementing agencies are the International Rescue Committee (IRC) (health, mental health, adult education, micro-enterprise), Don Bosco (vocational training) and World Vision (WV) (construction of shelters).

Food is distributed to those who have ration cards every fifteen days. The amount and contents of a 'food basket' is the same regardless of age and sex: 3.3 kg of wheat flour, 3.5 kg of maize, 0.6 kg of lentils, 0.6 kg of maize-soya blend flour, 0.45 kg of oil and 75 g of salt.[7] Vegetables are occasionally distributed. According to LWF/DWS, this 'basket' provides 2,100 kilocalories per person per day.[8] Education, health care and other services provided by agencies are free of charge. There are 27,000 pupils enrolled at pre-primary (kindergarten), primary and secondary schools.

5. Cambrenzy (1996).
6. LWF/DWS (1999).
7. The figures were shown to me by a Pari refugee who is the 'group leader' of Block 10.
8. LWF/DWS (1999: 9). Due to financial problems, the ration was reduced in August 2000 to 6.75 kg of maize, 0.6 kg of maize-soya flour and 75 g of salt, which amounts to 1,727 kilocalories per day. Wheat flour, pulses and cooking oil were not distributed. The August food distribution was met by stone-throwing by refugees, which had to be halted by police firing in the air ('UNHCR Press Briefing Note, 18 August 2000' (http://www.unhcr.ch/cgi-bin/texis/vtx/news/opendoc.htm?tbl=NEWS&page=home&id=3ae6b8274c). When I visited Kakuma in September, the ration was still cut down, a source of much grievance among refugees.

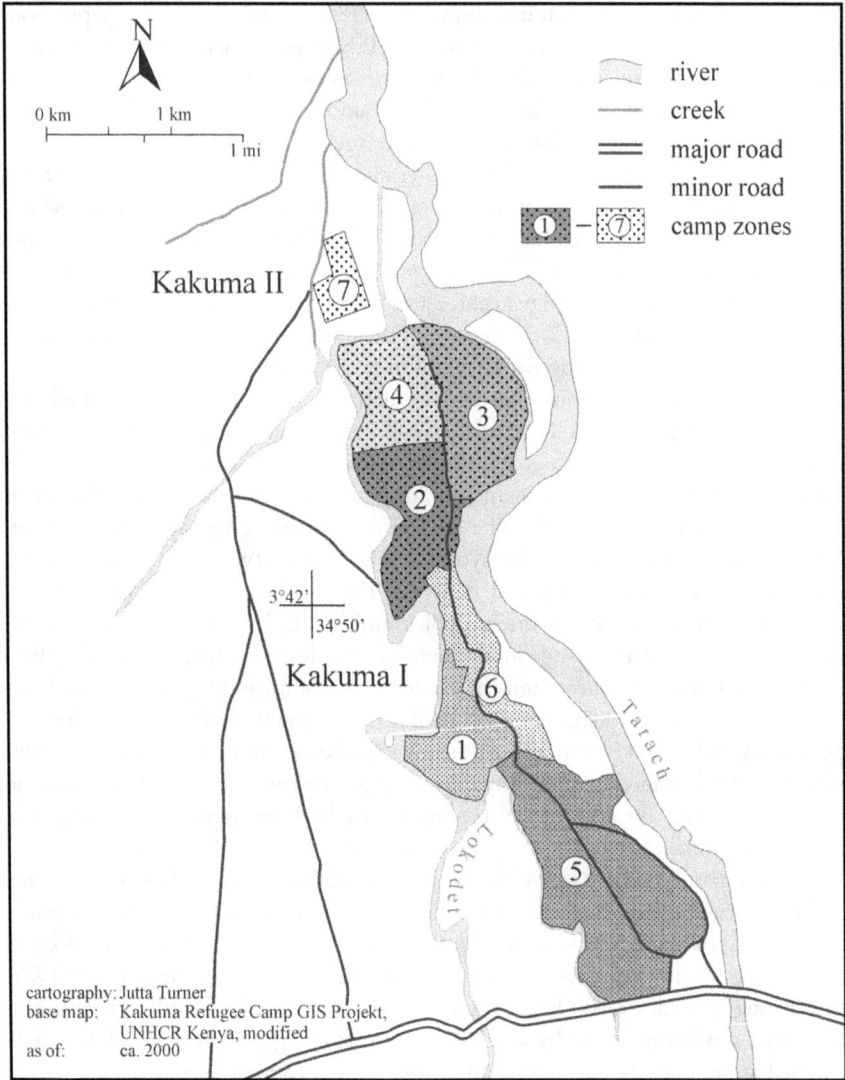

Map 10.1 Kakuma Refugee Camp

The Social Space of Kakuma

The natural environment is very arid, which makes food production activities aimed at enhancing refugees' self-reliance virtually impossible. It is hot, windy and dusty throughout the year. This is a hostile and unaccustomed climate for most of the refugees, with the possible exception of the Somali people. The local people are the pastoral Turkana, who have settlements in the vicinity of camp and frequent the camp for trade.

The Kakuma camp extends along the north-south axis (see Map 10.1). Both sides are demarcated by rivers, which are dry for most of the year. The length (north-south) is about 6 km and the width (east-west) is between 0.2 and 1.6 km. The northern end is where the two rivers merge, and the southern end is the tarmac road connecting Kakuma to Nairobi. Now this is called 'Kakuma I', as new extensions – 'Kakuma II and III' – have been set up to the north-east of Kakuma I in order to accommodate new arrivals. The UNHCR compound is located near the southern end. It also houses the offices of LWF/DWS, IRC and WV and the residences of their staff. Further south on the tarmac road is the Kenyan police station that is in charge of the security of the camp.

The social space of Kakuma refugees is in principle confined to the geographical boundaries of the camp. They hardly go out of it. On the other side of the rivers are Turkana settlements, but few refugees cross the dry riverbeds, although Turkana men and women do come to the camp. This means that the bush surrounding the camp is not utilized either for foraging or for collecting firewood, which, they say, is not allowed by the Turkana. After all, the land is dry and natural resources are very scarce. The only exceptions are some Dinka, who buy cattle, goats and sheep from Turkana in exchange for cash, bullets and firearms, and keep them with the original owners. In this way they establish a form of bond friendship and visit each other.[9]

The densely populated Kakuma camp[10] is not simply a refugee settlement, but a 'town'.[11] In fact, Kakuma camp is much more urban than Kakuma town, located next to the camp. It has a number of infrastructures for public purposes: kindergartens; primary schools, which have about 20,000 pupils; twenty-one secondary schools; four high schools; two technical colleges; a ninety-bed hospital and four clinics with a total capacity of 520,[12] a South African correspondence university, opened in 2000; 'multi-purpose community centres'; 'youth and resource centres'; 'women's multi-purpose centres'; and libraries. There are also many churches and mosques built by refugees. What is amazing are the commercial and trade activities carried out in open markets with stalls and in shops. Particularly impressive are the 'shopping centres', which stretch for more than one kilometre along two parallel main roads in the south of the camp. Both sides of the roads are full of kiosks selling a variety of commodities: they include butcheries, groceries, tea and coffee

---

9. Dr Itaru Ohta, personal communication. He has been doing research in Kakuma among the Turkana for twenty years, since well before the opening of Kakuma camp. A number of Pari unanimously say that they have no social relations with Turkana at all, although the Turkana do visit Pari households for trade and sometimes the Turkana eat there. They also mention that some Dinka have social relations with Turkana. According to Dr Ohta, those Turkana claim that they speak 'Dinka'. When he showed me a word list of this 'Dinka' language, interestingly it includes a number of Arabic and also Amharic words, presumably reflecting their stay in the refugee camps in Gambela, western Ethiopia.
10. In some blocks the population density is as high as 400 per hectare (Ferouse de Montclos and Mwangi Kagwanja 2000).
11. Ibid.
12. Ibid.: 210–11.

houses, bars and restaurants, hotels, satellite TV and video theatres and hair salons. There is even a place where international fax and telephone services are available. Although Sudanese refugees participate as petty traders, most of the business people are Ethiopians and Somalis.[13] Walking along these busy streets is quite an experience; you see different kinds of people including local Turkana men and women, and hear different languages – Swahili, Amharic, Oromo, Somali, Juba Arabic, a lingua franca in south Sudan, English and French. There is no doubt that Kakuma has emerged as the largest urban and trade centre not only in north-western Kenya but in a wider region, including parts of south-western Ethiopia, south-eastern Sudan and north-eastern Uganda.[14]

The projects of UNHCR and NGOs provide jobs not only for expatriates and Kenyans but also for about 2,000 refugees. The hospital employs seventy-eight refugees and primary schools employ 341. They are paid 'incentives', not salaries, varying between about 2,000 and 3,000 Kenyan shillings per month.[15] Manual labourers are paid something like 100 Kenyan shillings per day per person. The money paid contributes to sustaining the livelihood of refugees and the commercial activities in the camp.

Kakuma is an extraordinary refugee camp, which is multinational, multi-ethnic, multilingual and multi-religious. It is urban by nature with seemingly adequate infrastructures and programmes to foster a 'cosmopolitan' atmosphere and to build up a sort of civil society. The general trend, however, is not in that direction. I could find little evidence of the creation of significant new social ties crossing existing boundaries or of hybrid cultural practices. In this sense, Malkki is right when she argues that being refugees and being in exile are entirely different.[16] I wonder whether refugees in North-East Africa, no matter how many years they live as refugees, will ever produce an artist or a writer.

What we can observe in Kakuma is a fragmentation of identifications. Among the south Sudanese, instead of cultivating a Sudanese or south Sudanese identity, there has been an increasing move towards the politicization and fragmentation of identification. What seems to be closely associated with this is the residential pattern, which is strictly ordered by ethnicity and nationality. For administrative purposes, the camp is divided into seven 'zones' and about eighty 'blocks'. If we start from the south (Zone 5), the Nuer live in the southernmost area near the UNHCR compounds. Then come Ethiopians and Somalis, who developed the 'shopping centres' in their own residential areas. A small group of Congolese live next to the Somalis. Then, the northern part of Zone 5 and neighbouring Zones 1 and 6 are occupied by 'Equatorians', south Sudanese from Equatoria province. The northern part, Zones 2–4, is predominantly Dinka, while newly arrived Somalis occupy the northernmost part (Zone 7).

---

13. Ethiopians were the first to start businesses, followed by Somalis who were relocated from other camps and from Mombasa on the Indian coast (ibid.: 213).
14. Dr Ohta says that the commercial activities in the camp were even more flourishing a few years ago. It seems that the population decline of Ethiopian refugees is a major cause for this change.
15. Ferouse de Montclos and Mwangi Kagwanja (2000: 218).
16. Malkki (1995b).

The UNHCR administration classifies south Sudanese into three categories or 'tribes': Dinka, Nuer and 'Equatorians'. The south Sudan is composed of three administrative provinces or regions – Bahr al Ghazal, Upper Nile and Equatoria – and the three 'tribes' largely overlap with them: Dinka-Bahr al Ghazal, Nuer-Upper Nile and Equatorians-Equatoria, although the Bor section of Dinka live in Upper Nile. This classification may reflect political developments within south Sudan, although it is not certain to what extent the UN officers who adopted this classification were knowledgeable about them. In any case, central to these political developments are the split within the Sudan People's Liberation Army (SPLA) in 1991 and the ensuing inter-factional fighting. It took the form of inter-'tribal' conflict, namely between Dinka, who allegedly supported the SPLA-Torit (later SPLA-Mainstream) faction led by Dr John Garang, the founding Chairman and Commander-in-Chief and a Dinka from Bor, and Nuer, who rallied around the SPLA-Nasir (later SPLA-United, and then South Sudan Independent Movement, and finally the core parts joined the Khartoum government) led by Dr Riek Machar, a western Nuer. When they arrived in Kakuma in 1992, the Nuer refugees themselves requested that they wanted to stay away from the Dinka. The separation could not prevent conflicts between the two. There was a major clash between Nuer and Dinka in 1996, leaving eight dead and 148 seriously injured. In June 1997 they clashed again and more than 100 were injured. At the end of 1998 there were two hand-grenade explosions in Dinka and Nuer residential areas.

In Kakuma it is not only Dinka and Nuer that fight. This is a refugee camp where violent crimes at the individual level – robbery, rape and murder – and violent conflicts at the communal level are rampant. Security issues are the primary concern for the UNHCR.[17] Another very violent conflict happened in February 1999, this time between Dinka and Didinga. Dinka men came and attacked Didinga in their homes. Six were killed, 300 injured. Four hundred shelters (huts) were razed to the ground, making 6,500 homeless. They became refugees in their own refugee camp. The Didinga are an ethnic group in eastern Equatoria, whose administrative centre is Chukudum. There a Dinka contingent of the SPLA had been behaving like an occupying army, plundering and looting local Didinga. After the killing of the Didinga chief, they stood up and killed the Dinka commander and his bodyguards. They had to meet a massive retaliatory operation by the SPLA forces, which was contained only after an intervention by the SPLA leadership.

Dinka refugees themselves are by no means politically unified. The major divide is between the Dinka of Bahr al-Ghazal and the Dinka of Bor. They fought in December 1997, leaving 140 injured, in June 1998, leaving 155 injured, and again in April 2001. This time six were killed and 136 wounded. There are also reports of clashes between Sudanese and non-Sudanese refugees. In March 1998 Nuer attacked Ethiopians and twenty-nine were injured. In September 1998 Sudanese attacked Somalis, burning fifty-four shelters and shops. One was killed and thirty were injured.

---

17. Crisp (2000).

To bring peace among south Sudanese refugees became a major concern for the UNHCR. While a peace education programme was being implemented, a large-scale peace meeting was held in January 2000, attended not only by parties to the conflict but also by Kenyan authorities, SPLA representatives and Sudanese local NGOs. They agreed to resolve all these conflicts, and a peace and reconciliation celebration took place.

In a situation of ethnic hostility and conflict, the classification and labelling of ethnic groups are themselves a highly sensitive and political issue. In this regard, the very problem lies in the 'official' classification of Sudanese refugees into three major 'tribes', the Dinka, Nuer and 'Equatorians'. First, although we may say that the Dinka or the Nuer constitute an ethnic group, each is a configuration of various 'tribes' or 'sections', with considerable linguistic and cultural differences. Moreover, although many Dinka support the SPLA headed by Dr John Garang (a Dinka Bor) and many Nuer support the movement of Dr Riek Machar (a western Nuer), it is highly misleading to say that each is a politically unified and homogeneous group. Clashes between Dinka Bor and Dinka of Bahr al-Ghazal are a case in point. And, again, we may not presume that the Dinka are a unified group. Rather, the reality may be that political affiliation and positioning are an individual matter, not automatically defined because of a person's ethnicity. More importantly, while reports by the UNHCR narrate the conflicts in ethnic terms, we do not know who instigated the attacks and who actually took part in them.

The case of 'Equatorians' is more problematic. First of all, 'Equatoria' is the name of an administrative province and has no ethnic implication. The linguistic and cultural diversity of ethnic groups in the province is great, much greater than of those in the other two provinces, Upper Nile and Bahr al-Ghazal. In terms of contemporary politics, they are far from being homogeneous. While the Upper Nile and Bahr al-Ghazal are dominated in terms of population by the Nuer and Shilluk and by the Dinka, respectively, in the case of Equatoria, there is no such dominant ethnic group. Instead, there are numerous ethnic groups of middle and small population size, such as the Zande, Moru, Madi, Kuku, Bari, Lokoya, Lotuho (Latuka), Pari, Acholi and Didinga. This point also reminds us of the positions of ethnic minorities in the Upper Nile and Bahr al-Ghazal, who are completely undermined and not represented in the administrative terminology.

Going back to the issue of 'Equatorians', under the UN administration it seems that a new ethnic or 'tribal' identity of 'Equatorians' is being invented. Pari community leaders said that they do not feel comfortable with it because their existence is not recognized by the UN and NGOs and they cannot deal directly with them. I would argue that the creation of this new identity could have more profound effects. The three labels, Dinka, Nuer and 'Equatorians', can become politicized and be manipulated by political elites both in and out of the camp in a way that spoils the efforts to unify the south Sudanese. We should remember that the notion of 'Equatorians' first appeared as a political term in the early 1980s when the 'redivision' or 'decentralization' of the regional government was a highly debatable issue. A group of political elites from Equatoria, allied with the then President Nimeiri, argued for the 'redivision', claiming that that was a way to overcome the

'Dinka domination' in regional politics. We know that the 'Equatorian' identity was politicized and manipulated by certain political elites at both regional and national levels for their own advantage and thus the regional autonomy collapsed, which became a major cause of the present civil war. I am not certain whether the UNHCR officers are aware of this. We understand that for administrative purposes south Sudanese refugees should be classified in some way, but we need to be aware that the categorization of people itself is a sensitive issue, especially when, as is the case here, categories have already given historical and political connotations, and the people themselves do not necessarily agree with a given category.

In any case, now there are three parallel administrative structures for the Sudanese refugees: the Sudanese Refugee Community Administration (Dinka, Nuba and a few Shilluks), the South Sudanese Refugee Community Administration (mainly Nuer and Shilluks) and the Equatorian Refugee Community Administration. Each has its own elected chairman and executive committee and may deal directly with the UNHCR, although on their part the UNHCR is careful in handling the issue so as not to deepen the divisions among the south Sudanese.

As argued by Liisa Malkki, 'the refugee camp is a vital device of power'.[18] The spatial structure of Kakuma itself represents power relations among refugees and between refugees and the UNHCR and other NGOs. What I said of the 'cosmopolitan atmosphere' of the 'shopping centres' may exist only at the surface level. For most of the south Sudanese at home, traders are primarily 'foreigners' or 'outsiders': Arabized and Islamized Sudanese of 'red skin'. In Kakuma, although the traders are replaced by Ethiopians and Somalis, the same characteristics persist. Those traders and businessmen are not partners with whom they may have amicable relations. It is also doubtful to what extent refugees see those infrastructures as their own – I mean, institutions with such wonderful names as 'multi-purpose community centres', 'women's multi-purpose centres' and 'youth and resource centres'. In addition to them, in Kakuma town there is a 'Turkana-Refugees Friendship Hall', which was constructed by the UNHCR.

The most visible symbol of power is the UNHCR compound itself. It is surrounded by an iron fence and barbed wire. All staff and guests are required to stay in the compound after 6 p.m. for security reasons. The only entrance is a double gate, where refugees are questioned and body-checked by security personnel. The outside of the gate is always full of refugees who want to visit the offices for one reason or another. It is not easy to enter the compound. The inner part of the compound is demarcated by another fence, where there are quarters with a bar, a cafeteria and residences (where I stayed as a guest). No refugee is allowed to enter there. When I asked the reason for the regulation, the UNHCR Protection Officer (whose work is to protect refugees) said that a refugee might bring in a bomb. Such mistrust and fear do exist, and it is inevitable for the refugees to see the world inside the fence as external and authoritarian.

---

18. Malkki (1995b: 498).

Pari Refugees in Kakuma

The Pari are a Luo (Western Nilotic)-speaking people living at the foot of Jebel Lafon in Torit District, eastern Equatoria, Sudan. In the early 1980s their population was about 11,000. In 1985–86 a majority of young Pari men who were then in their twenties and thirties left home and joined the SPLA en masse. They were given military training in Ethiopia. Some of them returned home voluntarily and liberated Pari from the national army in April 1986. The 1991 split of the SPLA caused serious damage to the Pari, however. In February 1993 an inter-factional fight within the SPLA took place at Lafon, and the entire six villages originally constituting the ethnic community of Pari were completely burnt down. Since then the national army has been occupying the site and no Pari lives there. Now people are displaced and scattered. Many of them live in new settlements in the bush away from the former villages. Others stay at camps for the displaced along the Sudan-Uganda borders, in refugee camps and settlements in Kenya and Uganda or in urban centres – Juba and Khartoum – inside Sudan. And yet many men are still in the SPLA rank and file, deployed in various parts of Sudan. Only a few of them have succeeded in resettling or obtaining scholarships to study in North America.

There are about 800 Pari refugees in Kakuma, of whom 245 are primary school pupils (200 boys and forty-five girls) and sixteen are secondary school students (all boys, no girls). Many households are headed by women. Their husbands are either dead or absent. According to my estimation based on the figures of some age-sets, more than half of the Pari men who were between twenty and thirty-five years of age in 1985, the majority of whom joined the SPLA, have died. They are the ones who are supposed to be the core of the community, but they are a completely 'lost' generation and their wives and children have been suffering. Those who are in Kakuma are the survivors. They managed to come to Kakuma to survive and to give an education to their children.

There is a shared dissatisfaction about the life in Kakuma. I heard unanimous complaints. The food is not sufficient, and its quality is poor. It gets finished before the next distribution and people stay hungry for a few days. The water supply to meet the needs of cooking, washing and bathing is very limited and women have to line up for hours to get it. The distributed firewood is completely inadequate. The hospital and clinic do not prescribe enough or good medicine. Both men and women say that, in spite of these difficulties, they stay there because of education. But, here again, a grievance is that it is not easy for children to go to secondary schools and, even if they complete secondary school, there is little chance to proceed to higher education.

A widow narrated her own experiences when my age-mates and their wives got together drinking local *araki* and beer:

> Greetings to my co-wives and husbands. I do not have much to say. I will only talk of one thing. I have fourteen years outside Pari. I left Pari when Pibor was captured [by the SPLA]. From there I proceeded to Pachalla and Itangi [a refugee camp in western Ethiopia]. From Itang I went to Dimo

[another refugee camp]. I travelled only with a child from Pari. In Ethiopia I got some children. When the situation became bad [in 1991] I came back to Sudan. As I heard there was a very good school at New Cush, I went there. But the school was not good. So I came to Kakuma to put my boys in school here. In Kakuma there is no good life. My boy stopped at Primary 7. I ask myself why the Pari boys are failing. Are they idle? Or is it because of poor teaching? What is wrong? I prayed that God would help me with all the journey I made, so that my boys would do well at school. I want to propose to you, the group leader, that we shall work together for education. Now I have got a shelter, the only remaining problem is school. It would be useless if the children and I should go back to Pari, after having spent fourteen years outside, without any result. I think that the reason for the failure of children is that there is nobody to support them. I am asking you, Kurimoto and Ukach [the group leader], to look into this school problem.

Perhaps the most important aspect of the Kakuma refugee camp is the provision of educational and training facilities. Given the fact that the education system is virtually non-existent in many parts of south Sudan, Kakuma camp has become a crucial place to produce educated south Sudanese.

It is remarkable that, given the above conditions, Pari refugees are trying to maintain and reconstruct community ties and indigenous social institutions, namely their age system. My visit in February 2000 coincided with a historic event: 'Madir', a new generation comprised of two age-sets, whose members are now forty to forty-eight years old, was taking over power from the senior generation to become *mojomiji*, the ruling age grade in charge of political and juridical affairs of the community. I was particularly pleased because I am a Madir, member of the junior age-set, and I was to have the opportunity to celebrate the taking over with the surviving age-mates. This takes place every ten years, and that of Kakuma was a follow-up to the main ceremony at home, which had taken place a few months before. I heard that other Pari communities in Khartoum and in the displaced camps in Sudan would follow Kakuma. This suggests that there are ties connecting people in the diaspora and those at home. In the taking-over ceremony, the drums were captured by Madir, the new *mojomiji*, and people danced to the drums and drank and sang. In this case, the drums were made of barrels and buckets, not of wood, and proper costumes and ornaments were scarce. The procedures of the ceremony were very much simplified and, above all, the new *mojomiji* were very few. They were only six. Despite all these constraints, the ceremony was planned and performed.

During my second stay, my age-mates proposed organizing a dance to commemorate my visit. I was asked whether I wanted to make an invocation (*lam*), a ritual speech that is central to a dance organized by the *mojomiji*. I was more than pleased to exercise the right as a new *mojomiji* and accepted the offer. On the day of the dance, we, the seven *mojomiji*, were fully dressed and led the marching and dancing around the drums set on the open ground, joined also by elders and married women and girls. After a while, the invocation started. We were expected to preach to younger ones and women and to pray for the general welfare of the community.

There were three speakers, including me, and the speech by Ukongo, the second one, was quite impressive. The contents were very different from what I used to hear in the village. The following is the text of his speech, which I quote at length.

Our plan is not only for today and tomorrow but for our grandchildren who are not yet born. Our grandchildren should not suffer, and should not be poor like us. The problem of school is very important here. We came here as poor people. We ran away because of death. We came here so that we would not die. So why should we start quarrelling among ourselves, saying one is from this village and another is from that village? This is bad. We are here as Pari. This drum is for the Pari as a whole, not for a particular village. We, the Pari who are in Kakuma, brought the drum out because of the students. We want to show them a way. They should find a way in future without a problem. Those who are clever among them will bring a drum like this. We have gathered here because of the drum. Without a drum, there would be nothing called Pari. We are here as Pari because of the drum.

You students, you make me disappointed when I see you with my own eyes. Some want to go back home, although they came here to study. When you do not get food today, you say that you are rejected. I want to tell you that during the time when we were students in Juba, food was prepared after we went to school. We used to get food once a day at three o'clock in the afternoon. If we missed it, we did not eat until the next day. But food is not very important.

Life is yours. After you complete your studies, because of your knowledge and virtue, you will be somebody responsible. You will be responsible for feeding your children. Sometimes the food that you asked a woman to cook maybe delayed. But don't complain. I would like to suggest to you that you organize yourselves into a group of five. Each of you cook one day, and in this way rotate for fifteen days. Then you know also the exact amount of food you consume during these days. You have to go ahead in your studies, and I do not want somebody to leave here because of food. We want you to be the people of future. If I beat you because of your mistake, you may say that you are not wanted. It is not because you are not wanted. I do that because I want you to go ahead.

Now, you women, you also have a problem. Your name is spoilt. People come here from home, saying, 'I am going to my brother/sister', 'I am going to my mother's sister's son/daughter', 'I am going to my mother's brother's son/daughter', 'I am going to my father's son/daughter'. But after a while he goes back to the village from here, and says that people of Kakuma did not give him food, and I ran away because of hunger. Your name is really spoilt. My advice is that, when there is a little food, call everybody and put it together so that they share that little amount. You share food with them. Don't say, 'This is my home. I do not care for them.' Don't let them go. If they go, your name will be spoilt.

Next is the problem of water. We are thinking of leaving it to you [women], because you don't respect the way we arrange things. When we assign a man [to supervise water fetching], you say that he misuses his power for his own benefit. Now you try your own way. If you don't succeed, it is up to you. If you say that an arrangement can be done, let it be done.

Now, the sickness which is affecting small children, may it go out today! (Go out!) May it go together with the sun today! (Go!) Those children who got sickness, may they get up! (Get up!). May they get up! (Get up!)

Ukongo presented his view on the present conditions and his visions for the future. The notion of poverty was something that I had never heard about before. In the refugee camp they do not possess the means of food production and foraging that they used to possess at home, and they are more exposed to the world of affluence, represented by the UN agencies and international NGOs, so they have started to conceptualize their own situation as being impoverished. He saw education as the key to get out of it and urges students to study hard regardless of the harsh conditions. While preaching a new ideology, he also argued for the observance of indigenous values: unity of the Pari as a whole, symbolized by the drum, respect for kin, relatives and guests and the importance of sharing and generosity. The speech by Ukongo was revealing and convincing, modern and traditional at the same time.

Then came my turn. The new *mojomiji* are called *jo-geedo*, 'builders', and I extended the metaphor by saying, 'Now the old hut that we lived in was destroyed. We the new *mojomiji* will construct a new hut for all of us to live in.' Then I started the last part of my speech by using the conventional phrase, as Ukongo did, 'The person caught by sickness and lying down, may he get up!' I went on to invoke, 'May the Pari get up! It is not only the Pari that I want to get up. May all the Equatorians get up! May all the southerners get up!' These are unusual invocations, but the audience shouted after me, 'Get up!', while women were ululating. I used these phrases because I thought the unity and alliance among Equatorians and southerners are essential for the future, and because 'getting up' (*oo maal*) also means 'to develop'.

Ukongo joined the SPLA in 1985 after he had graduated from an intermediate school and got married. After that, he was deployed in different places in Sudan, and travelled a lot in Ethiopia and Uganda, sometimes with his wives and children, sometimes without them. Now he has been settled in Kakuma since 1999 with two wives and his children. His eldest son is now studying at a secondary school in a refugee settlement in northern Uganda. For them it is the first peaceful period in which they have been living together for many years. He is a man of writing. He gave me his handwritten autobiography (4,000 words) and asked me to print it so that his children could read it. I did it using a printer in the UNHCR compound. He also showed me two handwritten manuscripts: one is a romanized translation of part of the New Testaments into Pari; the other, to my surprise, is a copy in block letters of my paper on the Pari age system. As he was planning to leave Kakuma soon to return to the SPLA, which was said to have been preparing a major offensive in eastern Equatoria, I took his autobiography as a sort of last bequest. I also thought that his act of writing was a means of affirming his 'Pariness'.

An interesting point found in Ukongo's autobiography is that, whenever he reached a new place, what he did first was to look for a Pari. Undoubtedly social ties among the Pari are instrumental for their survival. On the other hand, ties beyond the ethnic boundaries may be weaker than those within. During my stay I spent most of the daytime inside the huts, and I witnessed only four occasions when non-Pari south Sudanese refugees were visiting their home, sitting with Pari men and women, although many of their neighbours are non-Pari 'Equatorians'. I would say that most of the daily social interactions are confined to within the Pari community.

Stick fights are an occasion to express Pari men's solidarity and manhood. At home it is fought between age-sets and villages. In Kakuma, Pari men fight other ethnic groups. It is triggered off by a simple quarrel or insult. They fought on four occasions in the past and, they are proud to say, they beat their enemies. They say that a reconciliation ceremony was held after each stick fight, attended by both ethnic community groups, and thus no hostility was left behind.

The unifying force was exemplified by the formation of the 'Pari Community Development Association'. During my first stay there were initiatives to organize it, and it was formed in April. Its existence is still nominal, but it has a chairman, deputy chairman, secretary and treasurer. According to the handwritten charter its main purpose is to promote education and to construct a library. This was modelled on the association of the Moru refugees, which constructed a church and a library. What seems to be crucial for the new Association's success is how to establish external ties and manipulate them for their advantage. Community leaders are well aware of this, but at the moment they have not yet found resources at their disposal.

It is notable that in a refugee camp an ethnic community is not simply reconstructed. For instance, among Pari refugees, new elements are being incorporated and the old system modified. The discourse of modernity on education and development is an example. Other cases are related to women and Christianity. Now women not only participate in but also speak at community meetings, which could not happen before. Meetings are opened by Christian prayers, which is also new. However, there is no Pari pastor in Kakuma, and they do not have their own church.

I may have put too much emphasis on the unity of the Pari community. When we look at the speech by Ukongo from a different perspective, it reveals another aspect. The very fact that he repeatedly emphasizes unity and generosity suggests that there is a danger. One is a division of community on a village basis. A village is a structural unit that has a considerable autonomy, although the largest village, Wiatuo, has a hereditary chief who is a rainmaker and who exercises power over all six villages. In the early 1980s there were various inter-village conflicts, as a result of which one village, Kor, a historical rival of Wiatuo, detached itself from the other five villages and became independent. In the early 1930s Wiatuo split into two, the breaking-away section forming its own village in order to receive their own share of relief food. Now that they no longer stay in one place at the foot of Jebel Lafon and are scattered in smaller settlements in the bush, communal annual ceremonies that used to be performed by all six villages are done separately. Many settlements stopped paying tribute to the rainmaker. Some Pari commented on the present situation and said that now people find these scattered settlements rather comfortable, as there is less threat of inter-village conflict. Therefore, Ukongo's emphasis on the unity of the Pari as a whole must be seen in this context. Also his speech tells us that some families are not generous enough to receive guests from home. Finally it suggests a conflicting relation between men and women. These issues will be pursued in future research.

In the multinational Kakuma, the Pari refugees live in a rather small social space. The momentum of their identification is centripetal. Certainly it helped them to survive and maintain their dignity. As a survival strategy, it may have disadvantages, and they may need a wider social space in which to breathe. Their survival depends on how they are able to manipulate different identifications – Pari, 'Equatorians', south Sudanese, Sudanese, refugees, Christians and possibly 'unaccompanied minors' and women – according to the context.

Chapter 11

Crossing Points: Journeys of Transformation on the Sudan–Ethiopian Border

Wendy James

We all have a fairly straightforward idea of what we might mean in principle by 'a journey'. In the normal sense it is defined by a beginning, a smooth and steady unfolding towards a goal and completion with the final arrival – for example, from Crewe to Birmingham, or even Nairobi to Arusha over an international border. I think of myself as the very same person throughout and those places as fixed – in social character more or less as in latitude and longitude. When we try to understand the social history of a disturbed zone, however, this is a hopelessly innocent image. When people have to leave one place because of insecurity and seek another, the direction and end purpose of the journey is not always clear. The crossing of a border may quite transform the 'identity' of individuals and groups, and the places they leave behind may be so changed by military events and socio-political history that they are really no longer the same places. Crossing points, and the ill-defined spaces between them, can become far more important than beginnings or endings in defining the journey. They can shape the personal sense of being, as well as the allegiances of people en route. A two-dimensional sketch map is too flat to capture the dramatic structure of such journeys. A single narrative account, unless very remarkable, is also too restricted a medium in which to capture the complexity of these journeys as collective and personal social transformations. At the very least, a whole set of narratives must be taken together, as Lisa Malkki and Julia Powles have attempted in their work on refugees from Burundi and Angola, respectively (Malkki 1995a; Powles 2000).

As anthropologists recognize, however, while personal narratives are important, they can only be one perspective on the radical character of political transformations emanating from national or global centres. Events on journeys of the kind so many refugees have undertaken are marked by points of fateful decision, enforced declarations of identity and allegiance and irreversible events of life and death. These turning points can leave deep marks on people's sense of who they are and with whom they have the kind of affinity in which they can invest their trust. There is very often no going back from these points where new commitment has to be made. It is at international border crossings that the most dramatic transformations are required, evoking for the anthropologist something of the character of classic 'rites

Map 11.1 Uduk home area and the Sudan-Ethiopia borderlands

of passage', including the moral and spiritual echoes of 'liminality' in those treks through the no-man's-land of the bush between states or zones of protection and the personal trauma of transformation on acceptance into a new place.

In this account I shall try to capture something of this transformative character of treks through the bush and 'crossing points', by considering a range of evidence from interviews with displaced people from the southern Blue Nile Province of Sudan, specifically Uduk speakers, at different stages in their complex history of movement since their villages were destroyed by the Sudanese army and allied militias in 1987 (for further details, see James 1994, 1996, 1997, 2000a, b, 2001, 2002, 2007). The most obvious cases are those of international crossings, which the core of the Uduk-speaking people have undergone six times since 1987. However, crossing the lines of civil war or between the spaces controlled by different local government regimes or warlords can also be like 'rites of passage' moulding life itself. The character of such crossings has, moreover, a historical or time dimension as well as a spatial one. Map 11.1 shows some of the places that have made up the patterns of Uduk movements and the Appendix gives a summary of the main political events that have affected the peoples of this border region since 1972.

Thinking over the eruption of population movements and memories of these movements on the Sudan/Ethiopian border, I have come to realize that, while I can try as a geographically minded scholar to plot these journeys on a map and date them, they do not take place in the homogeneous space of the map-maker or the steady time of the historian. Space and time are not here a neutral background against which the human reality of an individual or shared journey can be measured or remembered. They have a political shape, often a shape whose boundaries shift in place and time and whose internal character and boundaries are defined in highly formalized ways. In considering political space, I am drawing partly on Paul Dresch and his work on Yemen (Dresch 1989). The reputation and honour of politically powerful shaikhs there depend upon their ability to protect those who live within their orbit; there is a shape to the social 'space' over which they preside. What is clear to the inhabitants of the Sudan-Ethiopian border is that there are changing shapes of 'political space' over time – not linked so much these days to the cultivation of honour among nobility and local elites but rather to the shifting shapes of post-colonial national political economies. What is essential to such a concept of political space? An area where some kind of authority or security exists and extends protection. Such areas expand and contract over time, almost like weather systems. If they are stable, you don't notice, but, at their boundaries and zones of contact, you do – especially when you have to move or think it advisable to cross over, to move from the protection of one political space to another. This is because you have to negotiate the boundaries, a negotiation that can transform your sense of who you are in quite profound ways; for example, it may involve a declaration of your religion or even your 'ethnic' identity and political allegiance in ways that have never been imposed before. This crossing of boundaries can happen in geographical space, but also in historical time as one regime takes over from another and redefines the ground upon which individual and group rights, claims and obligations to authority are staked.

In considering boundaries, I am drawing on the landmark edited volume on *Border Identities* (Wilson and Donnan 1998). The various studies collected here illustrate powerfully how concepts of nation and person may be clearer at the borders of a country than anywhere else, for it is at the border that everyone is more conscious of national difference and alternative possibilities, along with the artificiality of the border line itself and the specific ease or difficulty that may be involved in the actual crossing. On the border, one knows that the character of places is made by history and is not fixed for ever. One knows that, when a place is overtaken by a succession of socio-political upheavals, a person's otherwise reasonable feeling of a continuing connection with a place may be broken. Beyond strategic claims to new kinds of allegiance and identity, which political anthropology has long recognized, there may be deeply personal problems for migrants or refugees themselves as to who they are and what places and even former patrons they can make claims upon. These are certainly real dilemmas for the multiply dislodged communities of the Sudan-Ethiopian border region.

Consider some of the places on Map 11.1. Chali, for example, was a central market village and former mission station (1938–64), which still counts as 'home' literally and 'symbolically' for many displaced Uduk speakers (James 1979, 1988). However, like its district headquarters of Kurmuk, it has changed hands several times between the Sudan government and the SPLA since the mid-1980s and been quite transformed by the recent waxing and waning of socio-political space in this border region (see the Appendix for a chronology of events). Even those places where most of the Uduk sought refuge after fleeing from their torched villages in 1987 have changed character several times. A place like Itang, within Ethiopia but near the frontier, has seen refugees coming and going under a succession of different regimes in both Ethiopia and Sudan, each time with a new story of need, seeking new political allies in the search for security and survival. Such a place becomes a very complex node in anyone's memory. Sometimes the UN is there, sometimes it's not; sometimes the Ethiopian government is in control, sometimes it is challenged by armed movements of one kind or another. The fortunes of a frontier settlement like Itang are sometimes driven by local events but are often entwined with events at some political centre, over which locals have little influence. When the socialist government of Mengistu fell in 1991, for example, the flow of refugees went into reverse; instead of coming into Ethiopia, Sudanese (and even some Ethiopian) people in Itang were desperate to leave and went downstream to Nasir. Some 20,000 Uduk, who had been in Itang for ten or eleven months, were among them; but, after less than a further year under the Nasir faction of the Sudan People's Liberation Army (SPLA), they sought to return to Itang. Itang in 1992 was a profoundly different place of refuge from Itang in 1990. For the Uduk who had to leave Sudan and seek protection there each time, it was ghastly, but for totally different reasons. On the first occasion they were fleeing with the SPLA from the Sudanese national forces and entering (in fact re-entering) the sympathetic territory of Mengistu's Ethiopia. On the second, they were fleeing from the SPLA itself (or rather from a local splinter of that movement) and entering the 'new' Ethiopia of the EPRDF, sympathetic no longer to the SPLA but rather to Khartoum. Even people who sit in a place like Itang through political changes of this kind experience a kind of 'crossing over' in time, sometimes a very visible and tangible transformation of social authority, as one local government regime displaces another.

Social and Religious Transformations: Boundaries and the Spaces in Between

In this section I give some examples of the transformative broundary-crossing experiences of the Uduk and the way they are remembered later. Many such crossings involve a registration of names (of self, father and father's father, regardless of local ways of naming), of ethnic group, age and nationality and of religion. People are obliged to declare how they fit into these categories of official identity, however fluid individual and group affiliations were before.

In the first example below, Martha Ahmed's account, given to me in 1993, refers at first in very general terms to the long trek through the bush and the way it transformed people into accepting the God of the Bible. The crucial point at which people's religion had formally been registered, however, was at the reception desk of the first refugee camp where they arrived after their initial border crossing from Chali to Assosa in 1987. Martha goes back in time, as it were, to give a graphic account of this actual declaration in the registration queue. She herself had been a Christian from childhood, but the man behind her sitting in line had not: his 'conversion', such as it was, happened at this point. Martha reminds us that the same questions were asked when they arrived back in Ethiopia at the Itang camp in 1990, and by this time she appears to think that the people really were sincere about it, as a result of surviving the terrible experiences of the journey itself. The intermediate sections of the long journey, when they were in between one patron or secure area and another were the times when most people died. During the whole period of the 1983–2005 Sudanese war, in the case of the people from the Blue Nile at least, most who died were not been killed by bullets or bombs but by hunger, disease and exhaustion in the 'liminal' times and spaces. At these times too, they have been especially vulnerable to spiritual experience, at least in retrospect, a retrospective that can take the form of dreams and certainties of religious conversion. This feeling of a reality that has been made known through the long journey in the wilderness can underpin a feeling of temporariness and unreality in the official rules and regulations of any particular refugee camp. The vulnerability of people to the spaces of the bush emerges vividly from my conversations with these refugees at different times and places over the last twenty years. A particularly strong theme is their wholesale acceptance of Christianity after many years of the complete absence of missionaries. Many individuals have painful memories of the trek, in which they lost kin and friends, sometimes literally dropping children under fire or having to leave older and weaker people behind to die under the trees. A feeling of inadequacy, even guilt, about such things can of course add to the appeal of a religion that promises personal salvation. The explicit declaration of allegiance often comes at a safe haven, where religion has to be registered in the UN books, and this can be the first such personal declaration. It can also be merely the most convenient thing to say, that one is a Christian. But there is a deeper rationale, as comes through clearly in Martha's account:

Well, in truth, it's because of us going through the bush. Those who had not believed before, they became like brand new people. From saying to themselves, 'I myself I am going to believe in the Word about Jesus, because I might die. And at that time of death, if I am going to die, then I will go to Heaven.' This is what people were thinking to themselves, trekking on through the bush as they have been for the last few years, as forest foundlings. They decided to recognize the God of the Bible.

A person who had not believed would say, 'I would like to believe in Jesus, because we've been so preoccupied going through the bush, and maybe we will all die, and where is your path?' Because you have read in the Word of Jesus there, if you have not been baptized, and you are not a believer, and your Liver here you have not yet given to the God of the Bible, at the time when you die, if you die in a state of sinfulness then there will be no way for you ...

When we arrived in Assosa, they asked us, all of us, 'Are you Christians?' and we said, in fact I myself said when they asked me, I said, 'Yes, I am a Christian, I am a child of Jesus and I belong to the church.' And then someone else, who was just living without faith, he was behind me [in the queue] and they asked him, 'Are you a Christian?' and he said, yes, he was a Christian. He spoke like this, but he had not been a Christian back home. It was there in Langkwai, in Assosa, that he came to believe. 'I am a Christian!' ... Because of us fleeing through the bush, coming as forest foundlings, my Liver told me that I should believe. That's why everyone now has taken the same path, they changed their minds as they came behind me. They changed their minds.

And even the diviners themselves, they now come to the church and say, 'I would like to be a believer!' They were not forced. It was just from themselves ...

Everybody was saying, 'I'm a person of Jesus, I'm a person of Jesus, I'm a Christian, I'm a Christian.' When we arrived at Assosa, everybody was saying this; when we got to Itang, and they asked us, everybody just said the same thing. We had not seen this before; maybe it was the God of the Bible who did this. Because in the old days, before people knew about the Word of God, it is true they didn't know the Word of God. They didn't know at all. It came when the white people arrived long ago, and told the people, saying, 'Jesus died for you, Jesus died for your sins.' They killed him, and he died, and stayed for three days, and then he went to Heaven. He will come back, and he will select those people who believe, and, if you don't believe, he will leave you on the ground.[1]

The religious theme of being saved or savable only by the God of the Bible is sometimes very explicit, both in ordinary discourse and in the sermons preached at

---

1. Recorded on film in the transit camp of Karmi, 1993, as part of the rushes of *Orphans of Passage* (James 1993), with acknowledgements to Granada Television.

Sunday services under the trees or in the simple wooden bench and thatched churches which the Uduk would build wherever they settled temporarily on the long journey. Suske is one of those who would talk freely of the Old Testament parallels. She was the first wife of Pastor Paul Rasha Angwo, who was himself stranded in Khartoum by the events of 1987 and who died there of illness a few years later. His second wife, Hannah, died in the Nor Deng camp near Nasir in 1991, and Suske explained to me the same year that they were living like the people of old. I asked, what people of old?

> Yes, we are living like the people of long ago, the Israelites [*'kwani Yahuda*]. That's why I say to people, because there are some people who don't understand yet, I say that we will wait, and then we will all believe. Like the Jews. And, when the time comes that everybody believes, our God will lead us and then look after us in our homeland. In our homeland where we shall live, but right now we are not going to find our homeland soon, no. We have to wait for the one God. We must wait. We must be patient. With our prayers, and we must believe in our Livers. I am a woman, a person of Jesus, from long ago. Now, I am just living my life, with God. I have not wanted to leave God, if I were to leave God, then my Liver would cry ... 'Why?' they would say. 'What is preventing you?' I would sit there, thinking, and everybody would say I am lost; pray for me, pray for me, pray for Suske a lot.
>
> And that's why I just said now, we have become like the ancient Jews, from living in the bush. The Jews used to live like this.[2]

Again and again, the promise of a secure place to stay, even accepted as a temporary one, led to disappointment, which was sometimes seen as betrayal, and to a fresh move elsewhere across the spaces. Accounts of their stay in Nasir make this clear.

The Search for Security: Fading Hopes in Nasir, 1991–92

Ethiopia now had a new government; it was sympathetic to Khartoum; one might have expected the SPLA to be able to consolidate its civilian base and protection of the displaced. In fact the SPLA split, and the Blue Nile refugees found themselves (like many other groups) in a very difficult situation in Nasir. The Uduk later reported all kinds of hardship and the lack of proper 'government', *hakuma* (Arabic), in Nasir. The texts I quote below reveal very clearly the search for protection, for a political space in which there is some kind of basic security of life and limb, and for the recognition of some kind of moral contract with the powers of the land. In listening to these accounts I was reminded of the tales I used to hear from the Koma inside Ethiopia in the 1970s, and of how at different times they decided to 'give their hands to the British' or to 'give their hands to Haile Selassie', that is, to accept their

---

2. Tape-recorded, August 1991, Nor Deng near Nasir.

suzerainty in return for protection. The concept of 'government' and its responsibilities was very prominent in the way the Uduk talked to me about their travails. Many of the older people still regarded the 'Sudan government' as their own government, which would call them back one day, when better leaders had taken charge and the government no longer wished to kill its own people. In Nasir, soldiers of the SPLA were feared because they cheated the people of their UN rations and took their crops. In at least one incident they killed an Uduk refugee, and what rankled most in the accounts I have heard was that the 'government' of Riek Machar did nothing about the case when they complained, even though the man who shot Musa was wearing 'the uniform of the government [*hakuma*]'. Dawud Hajar explained about their anger:

> They shot him dead, yes with a gun ... Kalashnikov. Then he remained there, in the water, and we in the clinic there, we heard about it the next morning, when he had been brought back. People brought him to the clinic, and we examined his body and found the bullet went in the groin here and as far as the lower buttock, there was an entry wound. We said, right, he has died in this way; so you should take him to the government [*sic*: Riek Machar] and report this to them. And we thought that something would be done. But nothing was done. Nothing happened at all.
>
> These things overwhelm us, we do not have the ability to do anything about it. What can we do? Because the man who caused our blood to be spilled out on the grass there has not been produced. This death we died out there was like the pointless death of some creature; just like an animal in fact, he was killed and he died just like a dog.
>
> Ethiopia is very good for us, as long as the government of Ethiopia does not do something against us. It is being very good to us. We want to stay until the government in Sudan improves and changes, and says it wants its people to go home and calls us to go back to our real home. And we will go this way [indicating the mountains] ... God has led us to Ethiopia, but, when the government of Sudan has improved, we will go home.[3]

The refugees knew that the Nasir SPLA was fighting the mainstream SPLA in late 1991, and that suspects were being killed in Nasir during the first half of 1992. A chief, Soya Bam, explained to me later that year, in the transit camp of Karmi, that things were worse now than in the nineteenth century, because at that time when people were scattered by slave raiders, at least his forebear Bamagud was able to bring the government (i.e. the Anglo-Egyptian Condominium) and gather his people together to live in peace.[4] I asked him rhetorically where good government was to be found today, and he replied: 'Where, yes indeed! There isn't one! The governments now just fight each other! It is only God who can help people. There is aggression everywhere.'

---

3. Tape-recorded in Karmi near Gambela, September 1992.
4. See James (1979: 34–59) for the way this was remembered in the 1960s.

Riek Machar did eventually allow the Uduk to move from the immediate vicinity of Nasir itself to Maiwut and Kigille, close to the border, in May 1992. But the rains failed, the hunger was great and, under a newly emergent faction, the people suddenly dashed back across the frontier to seek refuge in Itang again. Their acceptance in the new Ethiopia depended on an entirely different political scenario from the two previous occasions when they had sought safety in Mengistu's Ethiopia. Now, they claimed (with some justification) to be seeking safety from persecution by the SPLA itself. The following quotations from conversations I held with people in the transit camp at Karmi in August/September 1992 illustrate some of these themes. I should note that the term 'Dhamkin' applies primarily to the Nuer but also to northern Nilotic peoples in general. The main problem in crossing back to Itang at this time was the presence of the Gaajak Nuer on the border, who allowed the refugees through in the hope that their arrival would kick-start a new relief programme in the district, but at the same time looted all their possessions as they passed.

Dispossession: the Return to Itang and Gambela, June–July 1992

The refugees suffered further random thefts from soldiers and at least one killing in the Maiwut area. Others had been attacked 'by Nuer' as they tried to flee 'home' northwards to the Chali area, at that time held by the Sudan government. Peter Koma told me how he complained to the local SPLA commander, reminding him who the people were – that is, former allies of the SPLA itself. 'I asked him, why do you kill 'kwanim pa [the self-name of the Uduk], don't you know the 'kwanim pa, the ones who were in Nasir?'[5] But he also knew that the commander regarded them as both the same and yet not quite the same people, because the times had changed; they could not be trusted to return home without betraying the SPLA by revealing information about their movements. There was an American aid worker who occasionally visited Maiwut, and no attacks or thefts took place while he was there. Dale was remembered as a good man, but the UN otherwise scarcely had a presence there. Peter did not leave with the first group who decided clandestinely to leave again for Ethiopia in late June 1992, but only with a remnant group some weeks later, having got the personal blessing of the local commander.

Unfortunately, there was trouble at the border. A Nuer Gaajak militia stopped Peter and his group and held them up for three days. These soldiers were 'different from the SPLA'. They claimed to be Ethiopians, but in fact, he said, they were Sudanese people who said they were just working for the Ethiopians. These armed men took charge of the medicines Peter was carrying to help the people on the road, and took all their good clothes, saying they risked being looted further along the way by the Anuak.[6] These things should be left with them in their office and could be collected later. Peter refused. He told them that, if they were the real soldiers of the

---

5. This comment and others quoted in this paragraph were tape-recorded in Karmi in September 1992.
6. An alternative spelling of the Anwyaa of Dereje and Falge, this volume, and the Anyuak of Hutchinson, this volume. See Dereje Feyissa's chapter for discussion.

government they claimed to be, they should escort the refugees to the first police post in Ethiopia, at Tarpam outside Itang. On the fourth day they got up to go, and Peter was arrested and put in prison overnight. The next day they were allowed to go and given an escort (though the little radio that Dale had given Peter was taken 'for safe keeping'). There were sixty-two Uduk in the group, and a further number of other refugees coming with them from Nasir. After passing through a flooded area, they reached a dry place near the road to Itang. The escort then 'took hold of us by force, and grabbed things. Took blankets, clothes, all of them – good clothes they tore off people, then they left us and said go on then. Some people were quite naked, all their clothes were taken.' Peter reflected that there was no way to return to Sudan because they were still fighting the SPLA. 'There is no way to go back to Sudan. Because of this Islamic sharia: they do not want *masahin*, Christians ... So this is why we decided to come back to Ethiopia.'

New Life from No-man's-land

William Danga, his wife Pake and her son from a previous marriage came over the border with a few others later than the groups described above – in fact, the following year, early 1993, when the main core of the people had reached Bonga. He told me how Pake gave birth after they had left the Nasir-SPLA-controlled Sudan and been looted of their possessions on the way, but before they had arrived at the first Ethiopian police post/garrison at Tarpam, just outside Itang. The path was very wild, and they saw elephants along the way. They rested for the night on a little coloured sheet they still had, and Pake had her baby. A truck found them in the morning. William said:

> They asked us who we were, and wrote our names down, and gave us a card, and said we should ride in the truck. Pake was there with the baby in her arms. They said, 'Oh, was that baby born yesterday? What is his name?' I said, 'He doesn't have a name yet!' ... They said, 'No, that won't do, you should give him a name now. Give him a name and I can write it on the card.' I said, 'All right.' I said, 'Good. I will call him Dale.' So he wrote down his name as Dale.[7]

William spontaneously gave the name of the American worker with the World Food Programme who had been so helpful to them over the previous year in Nasir. Dale meanwhile, of course, had moved on.

In any case, while the refugees may think of 'the UN' as a single body, the section of the WFP attached to Operation Lifeline Sudan (Southern Sector) based in Nairobi was a very different creature from the UNHCR Addis Ababa, on whose mercy the Uduk refugees threw themselves in 1992. UNHCR Addis Ababa had at the time already closed down its refugee camps in the Gambela district and was very glad to see the end of the SPLA and (they hoped) the Sudanese problem in general. Though

7. Conversation tape-recorded in Bonga, November 1994.

they and the Ethiopian government reluctantly allowed the new wave of refugees to remain and, by 1993, had agreed to the establishment of Bonga and had installed officials at Tarpam, the child with the 'global' name Dale was truly born in a no-man's-land.

I have used a lot of narrative material to focus on the way that displaced communities face social and personal transformation, specifically at crossing points in the course of their search for secure political spaces where some kind of 'government' exists. In the liminal spaces and times between secure zones, people may be lost and fear may colour the memory of those parts of the journey (see James 1997, 2007). The open spaces where there is no international, national or even reliable guerrilla authority are the spaces where people are especially vulnerable to hunger and death. Later memories may trigger heavy thoughts, dreams and religious experience, even providing a rationale for firm statements about conversion. The concept of international authority in the form of the various UN agencies exists in the people's imagination and hopes, but in practice it is fractured and its presence ephemeral, especially when major boundaries are crossed. The Sudan-Ethiopian frontier was such a major boundary during the cold war and, because of the nature of the Sudanese civil war, continued to be so. But the open spaces are not without promise: the naming of baby Dale was a sign of hope, and occasionally there are stories of unexpected rescue. I conclude this essay with such a story.

Children lost in the course of the treks through no-man's-land are usually presumed dead or lost for ever, but occasionally they survive with the help of strangers and in most cases will grow up as members of another community. A considerable number of Uduk children were scattered by crossfire and lost at Itang during a violent incident in July 1992. Some were 'adopted' by Anuak families, and of these a few were either traced by the UN and reunited with their relatives in Bonga or have shown up of their own accord. Others may well by now be thoroughly assimilated as Anuak. However, memories are not always completely erased in these circumstances. What happens to the 'identity' of children who lose contact with their relatives during these journeys but do not forget 'who they are'? In Bonga in mid-2000 I heard the story of a young man who had tried to register as an Uduk refugee at another camp, Sherkole, set up in 1997 near Assosa. However, there was a problem because he did not speak the Uduk language, only Berta, and his cheeks bore the parallel scars that are a typical custom of the local Berta. The registration clerks, with their very bureaucratic ideas of 'ethnic identity', refused to allow him to register. He claimed that, as a child, he had been lost on the hazardous journey through the hills south from Assosa in 1990, falling down a ravine with his baby sister, as the people came under fire from supporters of the OLF. The two children survived, and were later taken in and brought up by a Berta family. The girl later became ill and died, while the boy went off and had various adventures in Ethiopia. He then decided to leave his adopted family and rejoin his own people, after the Sherkole camp was opened. The one proof he had of his original identity was the wooden backboard that Uduk children use to carry around their infant siblings, helping their mother look after the little ones. He had kept this. When he showed it to some of the Uduk in Sherkole, they recognized him as one of their own and worked out that his actual mother was now in Bonga. Contact was eventually made with her, and she confirmed that she had lost the two children in the trek through the mountains under attack and thought them dead. As I left Bonga, plans were afoot to bring the young man there to be reunited with his mother.

The refugee community in Bonga themselves had absorbed a few stray children in the course of the long journey south in 1990, including, for example, one boy from Jebel Maiak, who now speaks only Uduk. This transformation of lost foundlings into members of the community is not by any means a brand new phenomenon in this part of North-East Africa. I was so convinced by both the symbolic and the historical importance of this capacity for assimilating people from the bush and incorporating them into the world of the 'kwanim pa that I made it the key theme of my original book about the Uduk (James 1979). Today, despite the dramatic transformations of 'identity' caused by modern movements over so many boundaries and the fearful losses remembered from the open spaces, the basic motif of being able to convert strangers into kin remains a powerful one.

Following the 2005 Comprehensive Peace Agreement between Khartoum and the SPLM/A, the UNHCR was able to assist the return of many Sudanese. For the Uduk in Bonga and Sherkole, this took place through 2007–8. 'Home' turned out to be a very different place in harsh reality from fond remembrance: but that must be another story.

Appendix 11.1: Summary of main events affecting the Sudan/Ethiopian border and the people of the southern Blue Nile (especially Uduk speakers), 1974–2008 (See Map 11.1)

1972   Peace agreement in Sudan concludes seventeen years of civil war, which broke out just before the country became independent in 1956. Most of the struggle was limited to the south, and did not reach the Blue Nile province directly.

1974   The imperial government of Haile Selassie in Ethiopia is overthrown; socialist regime established, eventually to become known as the Derg and led by Mengistu Haile Mariam.

1983   Renewed outbreak of civil war in Sudan, starting in the far south. Foundation of the Sudan People's Liberation Movement/Army (SPLM/SPLA).

1985   First appearance of SPLA in Blue Nile, after establishing bases inside Ethiopia.

1986   Recruitment of some young men, including from the Uduk people. An attack was mounted on Chali, but it failed.

Jan.–March 1987 Counter-insurgency operations by Sudan government forces and allied militia. Flight of people, under SPLA escort, to Tsore in Assosa district. People remember it as a hunting-ground of their grandfathers, and then a place of former captivity in the nineteenth century (James 1996).

Oct.–Dec. 1987   SPLA took Kurmuk and Geissan. The Sudan government mounted a high-profile campaign, across the Middle East, to raise support for the retaking of Kurmuk and Geissan. Radio propaganda was intense (James 2000a). These places, among the first in northern Sudan to be taken by the SPLA, were represented as cities of the Arab homeland. The retaking took place just about a month later, after (as the SPLA claimed) the rebels had decided to leave with their captured equipment and loot. The refugees remained for the next couple of years in the Tsore refugee camp across the border in the Assosa district.

June 1989   Coup in Khartoum, foiling progress on peacemaking efforts by existing regime; present leader, Omar el Beshir, comes to power with backing of National Islamic Front. The war is intensified along with acceleration of plans to create an Islamic state in Sudan.

Nov. 1989   Second taking of Kurmuk by SPLA.

Dec. 1989–
Jan. 1990   With collapse of Berlin Wall and Soviet Union, the Mengistu government in Ethiopia is weakened.

  SPLA relinquishes Kurmuk for the second time; pursuit by Sudanese government and movements opposed to the Derg. Various accounts state that, as Kurmuk was being retaken by the Sudan government forces, the OLF, EPLF and TPLF moved in a coordinated way towards Assosa, in order to capture the pro-Mengistu garrison there. The SPLA base and the refugee camp at Tsore were more or less en route, and the SPLA was advised to get the refugees out of the way. Flight southwards through mountains of Assosa, being fired on by local elements of the OLF (remembered as raiders from a century ago) in a ravine. Flight back to Sudan at Yabus, bombing, flight to Bisho, remembered as a place of former refuge. However, shelling, long dispersal southwards under SPLA direction. On the way, they pass through home areas of various Koma groups; some recognize them from legends of the more distant past, and say, 'The Uduk are coming home!' They were originally told they could make a temporary settlement in the Koma area, at a site called Pagak right on the border, but this did not get approval from UNHCR or the Ethiopian government, and they had to go right down to Itang in Ethiopia, a refugee camp near Gambela, arriving June/July 1990.

May 1991   Fall of Mengistu's government in Ethiopia and collapse of SPLA presence in the country. All refugees obliged to move downstream to Nasir under SPLA direction. On the way, bombing by Sudanese air force.

August 1991	Split in SPLA; Nasir commanders attempt to keep Blue Nile refugees as evidence of need for international aid. Some escape and get back to Blue Nile homeland, but are vulnerable to attack by SPLA on the way.
June 1992	After suffering under regime of Nasir SPLA, seen as mainly composed of Nuer, the core of the Uduk population splits: a faction leads the dash back to Ethiopia, gets caught in crossfire of violence at Itang and flees to Gambela. They are permitted to move to a transit camp at Karmi (see Disappearing World film, James 1993).
January 1993	There was a violent outbreak in Karmi against the Nuer and other southern refugees who the Uduk felt had 'followed them there', and all the Blue Nile people were moved to a new UNHCR scheme at Bonga (James 1996, 1997). This place is remembered locally as a historical refuge of the Koma people from the raids of Jote Tullu of the Western Oromo. Remnants of the Uduk left behind in the Nasir region either move north to the Blue Nile, in Sudanese government-controlled territory, or cross over to join the scheme in Bonga during 1993–94.
1993–95	Sudanese government consulates were allowed to open in Assosa and in Gambela, and among other things their officials tried to tempt refugees back across the border (where agricultural labour was needed). Most preferred the relative safety of their exile in Ethiopia to the explosive situation in their home areas straddling the front line of the Sudanese civil war.
1996–97	SPLA advance into Blue Nile again, and civilians (all of whom have already been displaced at least once) again flee. In Jan. 1997 the SPLA took Kurmuk for the third time. UNHCR set up new reception centre at Sherkole, very close to former Tsore, in the Assosa district. This could be represented as the third time in remembered history that the Uduk people have been obliged to accept protection at this spot on the map. There is an Ethiopian government garrison right next to the refugee scheme, and the site overlooks the ravines and foothills below, including the Uduk homeland.
2001	The SPLA are still in Kurmuk. A number of Uduk people from Bonga have visited Sherkole and indeed crossed over to see Kurmuk. There are no problems at this border for them at the moment. A modest number of people, mostly from Sherkole, have quietly returned to live in their old homeland on the Sudan side, under the protection of the SPLA. Their position is obviously very precarious, as this region is officially still a part of

northern Sudan, and the commitment of the SPLA and the National Democratic Alliance to the long-term defence of such marginal areas is not at all clear.

2002   In Feb. 2002, the main splinter SPLA group, led by Riek Machar, decides to rejoin the mainstream SPLA, still led by John Garang. This splinter group was once known as the Nasir faction and most recently as the Sudan People's Defence Force. Its decision to reunite with the main southern-based movement comes after a period of several years cosying up to the Sudanese government, accepting their patronage and actually fighting against the SPLA. The Sudanese government intensifies its bombing campaign in the region of the oilfields in the south.

2003   Insurgency escalates in Sudan's western region of Darfur, provoking severe counter measures by Khartoum over the next several years.

2005   Comprehensive Peace Agreement signed on 9 January between the Khartoum government and the SPLM/A. Creation of new Government of National Unity; implementation of the Agreement to include special status for a new Blue Nile State within the north of Sudan, along with Abyei and Southern Kordofan/Nuba Hills; and a referendum in 2011 to be held in the South of the country, which will allow for the possibility of a vote for secession of the South. Conditions for implementation remain fragile, especially following the death of John Garang in a helicopter accident in June, 2005.

2007–8   UNHCR facilitates the return of Sudanese refugees from camps in several countries; the return of Uduk and other Blue Nile people in the Ethiopian camps is completed by mid-2008. The peace remains fragile and the outcome of the promised referendum very uncertain, especially for those whose homelands adjoin the North/South border. The conflict in Darfur continues, and peace remains very elusive there.

Bibliography

Abu-Manga, A. 1978. 'Fulani and Hausa Speech Communities in the Sudan: a Case Study of Maiurno in the Blue Nile Province'. MA dissertation, IAAS, University of Khartoum.
———— 1993. 'Resistance to the Western System of Education by the Early Migrant Community of Maiurno (Sudan)', in N. Alkali and R. Motem (eds), *Islam in Africa*. Ibadan: Spectrum Books, 117–34.
ᶜAli, A.I.M. 1972. *The British, the Slave Trade and Slavery in the Sudan, 1820–1881*. Khartoum: Khartoum University Press.
Alier, A. 1990. *The Southern Sudan: Too Many Agreements Dishonoured*. London: Ithaca Press.
Allen, T. 1994. 'Ethnicity and Tribalism on the Sudan-Uganda Border', in F. Katsuyoshi and J. Markakis (eds), *Ethnicity and Conflict in the Horn of Africa*. London: James Currey, 112–39.
Allen, T. and S. Heald 2004. 'HIV/AIDS Policy in Africa: What has Worked in Uganda and What has Failed in Botswana?'. *Journal of International Development* vol. 16: 1141–54.
Alnwick, D.J. 1985. 'Background to the Karamoja Famine', in C.P. Dodge and P.D. Wiebe (eds), *Crisis in Uganda: The Breakdown of Health Services*. Oxford: Pergamon Press, 127–41.
Amnesty International 2000. *Sudan: The Human Price of Oil*. London: Amnesty International.
———— 2004. *Sudan: At the Mercy of Killers – Destruction of Villages in Darfur*. London: Amnesty International.
Anderson, D. 1986. 'Stock Theft and Moral Economy in Colonial Kenya'. *Africa* vol. 56: 399–415.
Asad, T. 1970. *The Kababish Arabs: Power, Authority and Consent in a Nomadic Tribe*. London: C. Hurst and Company.
Bahru Zewde 1976. 'Relations Between Ethiopia and the Sudan on the Western Frontier, 1898–1935'. PhD dissertation, School of Oriental and African Studies, University of London.
———— 1987. 'An Overview and Assessment of Gambela Trade (1904–1935)'. *The International Journal of African Historical Studies* vol. 20, no. 1: 75–94.
Barber, J. 1968. *Imperial Frontier: A Study of the Relations Between the British and the Pastoral Tribes of North East Uganda*. Nairobi: East African Publishing House.
Barton, T. and G. Wamai 1994. *Equity and Vulnerability: A Situation Analysis of Women, Adolescents and Children in Uganda, 1994*. Kampala, Uganda: Government of Uganda and Uganda National Council for Children.
Bayart, J.-F. 1993. *The State in Africa: The Politics of the Belly*. London: Longman.
Beck, K. 1988. *Die Kawahla von Kordofan: Ökologische und ökonomische Strategien arabischer Nomaden im Sudan*. Wiesbaden: Franz Steiner Verlag.
———— 2004. 'Die Massaker in Darfur'. *Zeitschrift für Genozidforschung* vol. 5, no. 2: 52–80.
Behnke, R.H. and I. Scoones 1993. 'Rethinking Rangeland Ecology: Implications for Rangeland Management in Africa', in R.H. Behnke, I. Scoones and C. Kerven, (eds), *Range Ecology at Disequilibrium*. London: Overseas Development Institute, 1–30.
Behrend, H. 1999. *Alice Lakwena and the Holy Spirits: War in Northern Uganda, 1986–97*. Oxford: James Currey.

Birks, J.S. 1978. *Across the Savannas to Mecca: the Overland Pilgrimage Route from West Africa.* London: Hurst.

Bjørkelo, A. 1989. *Prelude to the Mahdiyya: Peasants and Traders in the Shendi Region, 1821–1885.* Cambridge: Cambridge University Press.

Boddy, J. 1989. *Wombs and Alien Spirits: Women, Men and the Zar Cult in Northern Sudan.* Madison, WI: University of Wisconsin Press.

———— 2002. 'Tacit Containment: Social Value, Embodiment, and Gender Practice in Northern Sudan', in S.J. Ellingston and M.C. Green (eds), *Religion and Sexuality in Cross-Cultural Perspective.* New York: Routledge, 187–221.

Boesen, E. 1999a. *Scham und Schönheit: Über Identität und Selbstvergewisserung bei den Fulbe Nordbenins.* Hamburg: LIT.

———— 1999b. 'Pulaaku: Sur la foulanité', in R. Botte, J. Schmitz and J. Boutrais (eds), *Figures Peules.* Paris: Karthala, 83–97.

Bokora Livestock Initiative (BOLI) 1998. *Training of Trainers Workshop for Community Animal Health Workers, 25–29 August.* Bokora Subcounty, Moroto District, Uganda.

Braukämper, U. 1971. 'Zur kulturhistorischen Bedeutung der Hirten-Ful für die Staatswesen des Zentralsudan'. *Paideuma* no. 17: 55–120.

———— 1992. *Migration und Ethnischer Wandel. Untersuchungen aus der Östlichen Sudanzone.* Studien zur Kulturkunde, 103. Stuttgart: Steiner.

Broch-Due, V. 1999. 'Remembered Cattle, Forgotten People: the Morality of Exchange and Exclusion of the Turkana Poor', in V. Broch-Due and D.M. Anderson (eds), *The Poor are not Us: Poverty and Pastoralism in Eastern Africa.* Oxford: James Currey, 50–88.

Bureau of Planning and Economic Development (BoPED) 1994. *Conservation Strategy of Gambela,* vol. I. Bureau of Planning and Economic Development, Gambela Regional Government.

Cambrenzy, L. 1996. *Kakuma: Monograph of a Refugee Camp, Preliminary Study.* Nairobi: UNHCR-ORSTOM

Central Statistics Authority 1995. *The 1994 Population and Housing Census of Ethiopia: Results for Gambela Region.* Addis Ababa: Central Statistics Authority.

Chabal, P. and J.-P. Daloz 1999. *Africa Works: Disorder as Political Instrument.* Oxford: James Currey.

Christian Aid 2001. *The Scorched Earth: Oil and War in Sudan.* London: Christian Aid.

Cisternino, M. 1979. 'Karamoja: The Human Zoo. The History of Planning for Karamoja with some Tentative Conterplanning [sic]'. PhD thesis, Post-Graduate School for Development Studies, Swansea University, Wales.

Civilian Protection Monitoring Team (CPMT) 2003. 'Humanitarian Comment on Pagak Area', 5 March, unpublished report.

———— 2004. 'Report of Investigation: No. 36, Fighting in the Shilluk Kingdom and Killing of Civilians', 11 April, unpublished report.

Clapham, C. 1996. *Africa and the International System.* Cambridge: Cambridge University Press.

Cole, D.C. and R. Huntington 1997. *Between a Swamp and a Hard Place.* Boston: Harvard Institute for International Development.

Collins, R.O. 1983. *Shadows in the Grass: Britain in the Southern Sudan, 1918–1956.* New Haven: Yale University Press.

———— 1971. *Land Beyond the Rivers: The Southern Sudan, 1898–1918.* New Haven: Yale University Press.

Comaroff, J. and J. L. Comaroff 1991. *Of Revelation and Revolution. Christianity, Colonialism, and Consciousness in South Africa,* Vol. I. Chicago, London: University of Chicago Press.

Crisp, J. 2000. 'A State of Insecurity: the Political Economy of Violence in Kenya's Refugee Camps'. *African Affairs* vol. 99: 601–32.

Crow, J.F. 1958. 'Some Possibilities for Measuring Selection Intensities in Man'. *Human Biology*, vol. 30: 1–13.

Cunnison, I. 1966. *Baggara Arabs: Power and the Lineage in a Sudanese Nomad Tribe*. Oxford: Clarendon Press.

Daly, M.W. 1986. *Empire on the Nile: the Anglo-Egyptian Sudan 1898–1934*. Cambridge: Cambridge University Press.

de Waal, A. 1997. *Famine Crimes: Politics and the Disaster Relief Industry in Africa*. London: James Currey.

———— 2004. 'Counter Insurgency on the Cheap'. *London Review of Books* (www.lrb.co.uk/ v26/n15/print/waalO1, accessed 10 February 2005).

———— 2005. 'Who are the Darfurians? Arab and African Identities, Violence and External Engagement'. *Contemporary Conflicts* (conconflicts.ssrc.org/Darfur/dewaal, accessed 10 February 2005).

Delaney, C. 1991. *The Seed and the Soil: Gender and Cosmology in Turkish Village Society*. Berkeley: University of California Press.

Delmet, C. 2000. 'Les Peul Nomades au Soudan', in Y. Diallo and G. Schlee (eds), *L'Ethnicité Peule dans des contextes nouveaux*. Paris: Karthala, 191–206.

Devlin, H. 1998. 'Patterns of Morbidity in Karamoja, Uganda, 1992–1996'. MA thesis, Department of Anthropology, Kansas University, Lawrence, KS.

Diallo, Y. and G. Schlee (eds) 2000. *L'ethnicité peule dans des contextes nouveaux*. Paris: Karthala.

Diallo, Y., M. Guichard and G. Schlee 2000. 'Quelques aspects comparatifs', in Y. Diallo and G. Schlee (eds), *L'ethnicité peule dans des contextes nouveaux*. Paris: Karthala, 225–55.

Donham, D. L.1999. *Marxist Modern: an Ethnographic History of the Ethiopian Revolution*. Berkeley: University of California Press.

———— 2001. 'Thinking Temporally or Modernizing Anthropology'. *American Anthropologist* vol. 103, no. 1: 134–49.

Doornbos, M. 2005. 'Transition and Legitimacy in African States: the Cases of Somalia and Uganda'. *FRIDE Working Paper* 17: 1–14.

Dresch, P. 1989. *Tribes, Government, and History in Yemen*. Oxford: Clarendon Press.

Duany, W. 1994. 'Report on the Processes of the Jikany-Lou Nuer Reconciliation Conference Held in Akobo, South Sudan, between July and October 1994'. Document.

Duffield, M. 1981. *Maiurno: Capitalism and Rural Life in Sudan*. London: Ithaca Press.

Dupire, M. 1962. *Peuls nomades: Étude descriptive des WoDaaBe du Sahel Nigérien*. Paris: Institut d'Ethnologie.

———— 1981. 'Réflexions sur l'ethnicité peule'. *Mémoires de la Société des Africanistes*, 2: 165–81.

Dyson-Hudson, N. 1966. *Karimojong Politics*. Oxford: Clarendon Press.

Dyson-Hudson, R. 1999. 'Turkana in Time Perspective', in M.A. Little and P.W. Leslie (eds), *Turkana Herders of the Dry Savanna. Ecology and Biobehavioural Response of Nomads to an Uncertain Environment*. Oxford: Oxford University Press, 25–40.

Ellis, J.E. and D.M. Swift 1988. 'Stability of African Pastoral Ecosystems: Alternate Paradigms and Implications for Development'. *Journal of Rangeland Management* vol. 41: 450–59.

Ellis, J.E., K. Galvin, J.T. McCabe and D.M. Swift 1987. *Pastoralism and Drought in Turkana District, Kenya. A Report to Norad*. Bellevue, CO: Development Systems Consultants.

Evans-Pritchard, E.E. 1940a. *The Nuer: A Description of the Modes of Livelihood and Political Institutions of a Nilotic people*. Oxford: Oxford University Press.

———— 1940b. *The Political System of the Anuak of the Anglo-Egyptian Sudan.* London: London School of Economics.

Fairhead, J. and M. Leach 1996. 'Rethinking the Forest-Savanna Mosaic', in M. Leach and R. Mearns (eds), *The Lie of the Land.* London: International African Institute, 105–21.

Falge, C. 2005. 'The Ethiopization of the Eastern Jikany Nuer from a Religious Perspective', in S. Kaplan, A.M. d'Alòs-Moner and E. Sokolinskaia (eds), *Workshop on the Historical and Anthropological Insights into the Missionary Activities in Ethiopia.* Berlin, Hamburg: LIT Verlag, 171–93.

Feaver, G. 1969. *From Status to Contract: A Biography of Sir Henry Maine. 1822–1888.* London: Longmans.

Ferouse de Montclos, M.-A. and P. Mwangi Kagwanja 2000. 'Refugee Camps or Cities? The Socio-economic Dynamics of the Dadaab and Kakuma Camps in Northern Kenya'. *Journal of Refugee Studies* vol. 13, no. 2: 205–22.

Fleay, M. 1996. 'Karamoja's District Team and Development Scheme', in D. Brown and M.V. Brown (eds), *Looking Back at the Ugandan Protectorate: Recollections of District Officers.* Dalkeith, West Australia: Douglas Brown, 20–25.

Fleisher, M.L. 1998. 'Cattle Raiding and its Correlates: the Cultural-Ecological Consequences of Market-Oriented Cattle Raiding among the Kuria of Tanzania'. *Human Ecology* vol. 26: 547–72.

———— 1999. 'Cattle Raiding and Household Demography among the Kuria of Tanzania'. *Africa* vol. 69: 238–55.

Fortes, M. 1969. *The Dynamics of Clanship among the Tallensi.* Oosterhout, N.B., The Netherlands: Anthropological Publications.

Fortes, M. and E.E. Evans-Pritchard (eds) 1940. *African Political Systems.* London: Oxford University Press.

Fox, R. 1993. *Reproduction and Succession.* New Brunswick, London: Transaction Publishers.

Franco, L. 1999. *Situation of Human Rights in the Sudan.* A/54/467, 14 October. Geneva: UN Commission on Human Rights.

Fukui, K. and J. Markakis (eds) 1994. *Ethnicity and Conflict in the Horn of Africa.* London: James Currey.

Gagnon, G. and J. Ryle, 2001. *Report of an Investigation into Oil Development, Conflict and Displacement in Western Upper Nile, Sudan.* Ottawa: Canadian Sudan Inter-Agency Reference Group.

Garfield, R. 2007. 'Violence and Victimization and Civilian Disarmament: The Case of Jonglei'. *Sudan Human Security Baseline Assessment Project, Working Paper 11.* Geneva: Small Arms Survey.

Gatdet, J. 2000. 'The Rise of Christianity among the Eastern Jikany Nuer (A Short History of the Rise of Christianity among the Eastern Jikany 1912–1993)'. Unpublished bachelor of theology, Mekane Yesus Theological Seminary, Addis Ababa.

Gertzel, C. 1976. 'Kingdoms, Districts and the Unitary State: Uganda 1945–1962', in D.A. Low and A. Smith (eds), *History of East Africa.* Oxford: Clarendon Press, 65–108.

Gilkes, P. 1999. *Ethiopia-Perspectives of Conflict 1991–1999.* Swiss Peace Foundation, Institute for Conflict Resolution and SDC, Federal Department of Foreign Affairs. (http.//www.swisspeace.ch/typo3/fileadmin/user_upload/pdf/FAST/CRPs/1999/Ethiopia-Perspectives_on_Conflict_1999.pdf).

GOU (Government of Uganda) 1994a. 'Cattle-rustling: National and International Dimension', in *All Eyes and National Attention to Karamoja. Proceedings of a National Conference on Strategies for Peace and Sustainable Development for Karamoja and*

Neighbouring Districts. Kampala, Uganda, 18–22 July. Kampala, Uganda: Ministries of State for Karamoja and for Presidential Affairs, 83–105.

——— 1994b. 'Immediate Challenges' in *All Eyes and National Attention to Karamoja. Proceedings of a National Conference on Strategies for Peace and Sustainable Development for Karamoja and Neighbouring Districts. Kampala, Uganda, 18–22 July.* Kampala, Uganda: Ministries of State for Karamoja and for Presidential Affairs, 139–75.

Gray, S.J. 2000. 'The Memory of Loss: Ecological Politics, Local History, and the Evolution of Karimojong Violence'. *Human Organization* vol. 59: 401–18.

Gray, S.J. and H.A. Akol 2000. 'Reproductive Histories and Life History of Bokora and Matheniko Karimojong Women of Northeast Uganda'. *American Journal of Physical Anthropology Supplement*, vol. 30: 165.

Gray, S.J., P.W. Leslie and H.A. Akol 2002. 'Uncertain Disaster: Environmental Instability, Colonial Policy, and the Resilience of East African Pastoralist Systems', in W.R. Leonard and M. H. Crawford (eds), *Human Biology of Pastoral Populations.* Cambridge: Cambridge University Press, 99–130.

Gray, S., M. Sundal, B. Wiebusch, M.A. Little, P.W. Leslie and I.L. Pike 2003. 'Cattle Raiding, Cultural Survival, and Adaptability of East African Pastoralists'. *Current Anthropology*, vol. 44 (Suppl.): S3–S30.

Gray, S.J., B. Wiebusch and H.A. Akol 2004. 'Cross Sectional Growth of Pastoralist Karimojong and Turkana Children'. *American Journal of Physical Anthropology* vol. 125, no. 2: 193–202.

Guichard, M. 1996. '"Les Fulɓe du Borgou n'ont vaincu personne": de la culture politique d'une minorité ethnique béninoise'. Doctoral thesis, University of Bielefeld.

——— 1998. 'Du discours sur la faiblesse du pouvoir fulbe', in E. Boesen, C. Hardung and R. Kuba (eds), *Regards sur le Borgou. Pouvoir et altérité dans une région ouest-africaine.* Paris: L'Harmattan, 185–202.

——— 2000. 'L'étrangeté comme code de communication interethnique: Les relations entre agropasteurs peuls et paysans bariba du Borgou (Nord-Bénin)', in Y. Diallo and G. Schlee (eds), *L'ethnicité peule dans des contextes nouveaux.* Paris: Karthala, 93–127.

Hall, S. 1990. 'Cultural Identity and Diaspora', in J. Rutherford (ed.), *Identity: Community, Culture, Difference.* London: Lawrence and Wishart, 222–37.

Harker, J. 2000. *Human Security in Sudan: The Report of a Canadian Assessment Mission.* Ottawa: Ministry of Foreign Affairs.

Hendrickson, D., J. Armon and R. Mearns 1998. 'The Changing Nature of Conflict and Famine Vulnerability: the Case of Livestock Raiding in Turkana District, Kenya'. *Disasters* vol. 22, no. 3: 185–99.

Hollis, A.C. 1910. 'A Note on the Masai System of Relationship and Other Matters Concerned Therewith'. *Journal of the Royal Anthropological Institute* vol. 40: 473–82.

Holt, P.M. and M.W. Daly 2000. *A History of the Sudan: From the Coming of Islam to the Present Day.* London: Longman.

Holy, L. 1991. *Religion and Custom in a Muslim Society: The Berti of Sudan.* Cambridge: Cambridge University Press.

——— 1998. *Anthropological Perspectives on Kinship.* London: Pluto Press.

Human Rights Watch 1999. *Famine in Sudan, 1998: The Human Rights Causes.* Washington, DC: Human Rights Watch.

——— 2002. 'LRA Conflict in Northern Uganda and Southern Sudan'. (http://www.hrw.org/press/2002/10/uganda1029-bck.htm).

——— 2004. *Sudan: Darfur Destroyed: Ethnic Cleansing by Government and Militia Forces in Western Sudan.* Washington, DC: Human Rights Watch.

Hutchinson, S.E. 1996. *Nuer Dilemmas. Coping with Money, War, and the State.* Berkeley, Los Angeles, London: University of California Press.

—— 1998. 'Death, Memory and the Politics of Legitimation: Nuer Experiences of the Continuing Second Sudanese Civil War', in R. Werbner (ed.), *Memory and the Postcolony: African Anthropology and the Critique of Power.* London: Zed Books, 58–70.

—— 2000. 'Nuer Ethnicity Militarized'. *Anthropology Today* vol. 16, no. 4: 6–13.

—— 2001. 'A Curse from God? Political and Religious Dimensions of the Post-1991 Rise of Ethnic Violence in South Sudan'. *Journal of Modern African Studies* vol. 39, no. 4: 307–31.

Hutchinson, S. E. and J.M. Jok 2002. 'Gendered Violence and the Militarization of Ethnicity: a Case Study from South Sudan', in R. Werbner (ed.), *Postcolonial Subjectivities in Africa.* London: Zed Books, 84–108.

Iliffe, J. 1979. *A Modern History of Tanganyika.* Cambridge: Cambridge University Press.

International Crisis Group 2002. *God, Oil and Country: Changing the Logic of War in Sudan, Africa Report* no. 39. Washington DC: International Crisis Group Press.

Jal, G. 1987. 'The History of Jikany Nuer before 1920'. PhD dissertation, History Department, School of Oriental and African Studies, University of London.

James, W. 1977. 'The Funj Mystique: Approaches to a Problem of Sudan History', in R.K. Jain (ed.), *Text and Context: The Social Anthropology of Tradition.* Philadelphia: ISHI, 95–133.

—— 1979. *'Kwanim Pa: The Making of the Uduk People. An Ethnographic Study of Survival in the Sudan-Ethiopian Borderlands.* Oxford: Clarendon Press.

—— 1988. *The Listening Ebony: Moral Knowledge, Religion and Power among the Uduk of Sudan.* Oxford: Clarendon Press (reissued in 1999 with a new Preface).

—— 1993. Documentary film *Orphans of Passage*, in the series 'Disappearing World: War', Granada Television, UK. Director: Bruce MacDonald. Set in the transit camp of Karmi.

—— 1994. 'Civil War and Ethnic Visibility: the Uduk on the Sudan-Ethiopia Border', in Fukui, K. and J. Markakis (eds), *Ethnicity and Conflict in the Horn of Africa.* London: James Currey, 140–64.

—— 1996. 'Uduk Resettlement: Dreams and Realities', in T. Allen (ed.), *In Search of Cool Ground: War, Flight and Homecoming in Northeast Africa.* London: James Currey, 182–202.

—— 1997. 'The Names of Fear: History, Memory and the Ethnography of Feeling among Uduk Refugees'. *Journal of the Royal Anthropological Institute* NS, 3: 115–31.

—— 2000a. 'The Multiple Voices of Sudanese Airspace' in R. Fardon and G. Furniss (eds), *African Broadcast Cultures.* Oxford: James Currey, 198–215.

—— 2000b. 'Beyond the First Encounter: Transformations of "the Field" in North East Africa', in P. Dresch, W. James and D. Parkin (eds), *Anthropologists in a Wider World: Essays on Field Research.* Oxford: Berghahn, 69–90.

—— 2000c. *Community Services for Sudanese Refugees in Western Ethiopia, Working Proposals for Bonga and Sherkole.* Unpublished research report for the UNHCR.

—— 2001 'Landscapes of War: Imagined and Inscribed'. Distinguished Africanist Lecture 2001, Institute for African Studies, Emory University, Atlanta, Georgia (http://www.emory.edu/COLLEGE/IAS).

—— 2002. 'No Place to Hide: Flag-Waving on the Western Frontier', in W. James, D.L. Donham, E. Kurimoto and A. Triulzi (eds), *Remapping Ethiopia: Socialism and After.* Oxford: James Currey, 259–75.

—— 2007. *War and Survival in Sudan's Frontierlands: Voices from the Blue Nile.* Oxford: Oxford University Press (www.voicesfromthebluenile.org).

James, W. and D.H. Johnson 1988. 'Introductory Essay: On "Native" Christianity', in W. James and D.H. Johnson (eds), *Vernacular Christianity*. JASO Occasional Papers No. 7. New York: Lilian Barber Press, 1–14.

James, W., D.L. Donham, E. Kurimoto and A. Triulzi (eds) 2002. *Remapping Ethiopia: Socialism and After*. Oxford: James Currey.

Johnson, D.H. 1982. 'Tribal Boundaries and Border Wars: Nuer-Dinka Relations in the Sobat and Zaraf Valleys, c. 1860–1976'. *Journal of African History* vol. 23, no. 2: 183–203.

———— 1986. 'On the Nilotic Frontier: Imperial Ethiopia in the Southern Sudan, 1898–1936', in D.L. Donham and W. James (eds), *The Southern Marches of Imperial Ethiopia: Essays in History and Social Anthropology*. Cambridge: Cambridge University Press, 219–45.

———— 1989. 'Political Ecology in the Upper Nile: the Twentieth Century Expansion of the Pastoral "Common Economy"'. *Journal of African History* vol. 30, no. 3: 463–86.

———— 1992. 'Recruitment and Entrapment in Private Slave Armies: the Structure of the *Zara'ib* in the Southern Sudan', in E. Savage (ed.), *The Human Commodity: Perspectives on the Trans-Saharan Slave Trade*. London: Frank Cass, 162–73.

———— 1994. *Nuer Prophets*. Oxford: Clarendon Press.

———— 1996. 'Increasing the Trauma of Return: an Assessment of the UN's Emergency Response to the Evacuation of the Sudanese Refugee Camps in Ethiopia, 1991', in T. Allen (ed.), *In Search of Cool Ground: Displacement and Homecoming in Northeast Africa*. London: James Currey. 171–81.

———— 1998. 'The Sudan People's Liberation Army and the Problem of Factionalism', in C. Clapham (ed.), *African Guerrillas*. Oxford: James Currey. 53–72.

———— 2001. 'The Nuer Civil War', in M.-B. Johannsen and N. Kastfelt (eds), *Sudanese Society in the Context of Civil War*. Copenhagen: North/South Priority Research Area, 3–27.

———— 2003. *The Root Causes of Sudan's Civil Wars*. Oxford: James Currey.

Jok, J.M. and S.E. Hutchinson 1999. 'Sudan's Prolonged Civil War and the Militarization of Nuer and Dinka Ethnic Idenities'. *African Studies Review* vol. 42, no. 2: 125–45.

Kaldor, M. 1999. *New and Old Wars: Organised Violence in a Global Era*. Cambridge: Polity Press.

Kaplan, S. 2005. 'Themes and Methods in the Study of Conversion in Ethiopia', in S. Kaplan, A.M. d'Alòs-Moner and M. Sokolinskaia (eds), *Workshop on the Historical and Anthropological Insights into the Missionary Activities in Ethiopia*. Berlin, Hamburg: LIT Verlag, 107–22.

Kapteijns, L. and J. Spaulding 1982. 'Precolonial Trade between States in the Eastern Sudan, ca 1700–ca 1900'. *African Economic History* vol. 11: 29–62.

Karugire S.R. 1980. *A Political History of Uganda*. London: Heinemann.

Kelly, R.C. 1985. *The Nuer Conquest: The Structure and Development of an Expansionist System*. Ann Arbor: University of Michigan Press.

Klumpp, D. and C. Kratz 1993. 'Aesthetics, Expertise and Ethnicity: Okiek and Maasai Perspectives on Personal Ornament' in T. Spear and R. Waller (eds), *Being Maasai: Ethnicity and Identity in East Africa*. Oxford: James Currey, 195–222.

Knighton, B. 2003. 'The State as Raider Among the Karamojong: "Where There are No Guns, They Use the Threat of Guns"'. *Africa* vol. 73, no. 3: 427–55.

Kronenberg, A. and W. Kronenberg 1965. 'Parallel Cousin Marriage in Mediaeval and Modern Nubia, Part 1'. *Kush* vol. 13: 241–60.

Kuper, A. 1982. *Wives for Cattle: Bridewealth and Marriage in Southern Africa*. London: Routledge and Kegan Paul.

———— 1988. *The Invention of Primitive Society*. London: Routledge.

Kurimoto, E. 1996. 'People of the River: Subsistence Economy of the Anywaa (Anuak) of Western Ethiopia', in S. Sato and E. Kurimoto (eds), *Essays in Northeast African Studies*. Osaka: National Museum of Ethnology, 29–57.

———— 1997. 'Politicization of Ethnicity in Gambela Region,' in K. Fukui, E. Kurimoto and M. Shigeta (eds), *Ethiopia in Broader Perspective: Papers of the 13th International Conference of Ethiopian Studies*, vol. 2. Kyoto: Shokado Book Sellers, 798–815.

———— 1998. 'Anywaa-Nuer Conflicts in Gambella Region, Jan.–Feb. 1998'. Confidential note, 24 February, Addis Ababa.

———— 2000. 'The Significance Patrilineal Descent in Changing Anywaa (Anuak) Society'. Paper given at the 14th International Conference of Ethiopian Studies. Addis Ababa.

———— 2001. 'Capturing Modernity among the Anywaa of Western Ethiopia', in E. Kurimoto (ed.), *Rewriting Africa: Towards Renaissance of Colapse*. Osaka: Japan Centre for African Studies.

———— 2002. 'Fear and Anger: Female Versus Male Narratives Among the Anywaa', in W. James, D.L. Donham, E. Kurimoto and A. Triulzi, (eds), *Remapping Ethiopia: Socialism and After*. Oxford: James Currey, 219–38.

Kwacakworo 1994. 'Notes on the Fighting in the Upper Nile between the Lou- and the Jikany-sections of the Nuer in 1993 and 1994'. Document. Nairobi: United Nations, Operation Lifeline Sudan.

Lamphear, J. 1976. *The Traditional History of the Jie of Uganda*. Oxford: Clarendon Press.

———— 1992. *The Scattering Time: Turkana Responses to Colonial Rule*. Oxford: Clarendon Press.

———— 1998. 'Brothers in Arms: Military Aspects of East African Age-class Systems in Historical Perspective', in E. Kurimoto and S. Simonse (eds), *Conflict, Age and Power in North East Africa*. Oxford: James Currey, 79–97.

Lang, H. 1977. *Exogamie und interner Krieg in Gesellschaften ohne Zentralgewalt*. Hohenschäftlarn: Kommissionsverlag Klaus Renner.

Lawrence, P. 1967. *Road Belong Cargo*. Manchester: University of Manchester Press.

Lesch, A.M. 1998. *The Sudan: Contested National Identities*. Bloomington: Indiana University Press.

Leslie, P.W., R. Dyson-Hudson and P.L. Fry 1999. 'Population Replacement and Persistence', in M.A. Little and P.W. Leslie (eds), *Turkana Herders of the Dry Savanna. Ecology and Biobehavioural Response of Nomads to an Uncertain Environment*. Oxford: Oxford University Press, 281–301.

Levine, I. 1996. 'Note for the Record, 13.4.96, Meeting with Dr. Timothy Tutlam of RASS'. Document. Nairobi: United Nations, Operation Lifeline Sudan.

Lévi-Strauss, C. 1969 [1949]. *The Elementary Structures of Kinship*. Boston: Beacon Press.

Lewis, I.M. 1999 [1961]. *A Pastoral Democracy: a Study of Pastoralism and Politics among the Northern Somali of the Horn of Africa*. Hamburg, Oxford: LIT Verlag, James Currey.

Little, M.A., R. Dyson-Hudson, N. Dyson-Hudson and N.L. Winterbauer 1999. 'Environmental Variations in the South Turkana Ecosystem', in M.A. Little and P.W. Leslie (eds), *Turkana Herders of the Dry Savanna: Ecology and Biobehavioural Response of Nomads to an Uncertain Environment*. Oxford: Oxford University Press, 317–30.

Lutheran World Federation/Department for World Service (LWF/DWS) 1999. 'Kenya/Sudan Programme, Annual Report 1999'.

MacMichael, H.A. 1922. *A History of the Arabs in the Sudan*, vols 1–2. Cambridge: Cambridge University Press.

———— 1954. *The Sudan*. London: Ernest Benn.

Maine, H.S. 1986 [1864]. *Ancient Law.* Tuscon: University of Arizona Press.

Majok, A.A. and C.W. Schwabe 1996. *Development among Africa's Migratory Pastoralists.* Westport, CT: Bergin and Garvey.

Malkki, Liisa 1995a. *Purity in Exile: Violence, Memory, and National Cosmology among Hutu Refugees in Tanzania.* Chicago: University of Chicago Press.

——— 1995b. 'Refugees and Exile: From "Refugee Studies" to the National Order of Things'. *Annual Review of Anthropology* vol. 24: 495–523.

Mamdani, M. 1976. *Politics and Class Formation in Uganda.* London: Monthly Review Press.

McCabe, J.T. 1990. 'Turkana Pastoralism: a Case Against the Tragedy of the Commons'. *Human Ecology* vol. 18, no. 1: 81–103.

Meggit, M.J. 1965. *The Lineage System of the Mae-Enga of New Guinea.* Edinburgh: Oliver and Boyd.

Meillassoux, C. 1991. *Maidens, Meal and Money: Capitalism and the Domestic Community.* Cambridge: Cambridge University Press.

Merker, H. 1910. *Die Masai.* Berlin: Dietrich Reimer.

Mirzeler, M. and C. Young 2000. 'Pastoral Politics in the Northeast Periphery in Uganda: AK-47 as Change Agent'. *Journal of Modern African Studies* vol. 38, no. 3: 407–29.

Mutibwa, P. 1992. *Uganda since Independence: A Story of Unfulfilled Hope.* London: Hurst.

Nasr, A.A. 1980. *Mai Wurno of the Blue Nile: A Study of an Oral Biography.* Khartoum: Khartoum University Press.

Nathan, M.A., E.M. Fratkin and E.A. Roth 1996. 'Sedentism and Child Health among Rendille Pastoralists of Northern Kenya'. *Social Science and Medicine* vol. 43, no. 4: 503–15.

Nyaba, P.A. 1997. *The Politics of Liberation in South Sudan: An Insider's View.* Kampala: Fountain Publishers.

O'Connor, A. 1988. 'Uganda: the Spatial Dimension', in H.B. Hansen and M. Twaddle (eds), *Uganda Now: Between Decay and Development.* London: James Currey, 83–94.

Ocan, C.E. 1992. *Pastoralist Crisis in North-Eastern Uganda: The Changing Significance of Cattle Raids.* Kampala, Uganda: Centre for Basic Research.

Okudi, B. 1992. *Causes and Effects of the 1980 Famine in Karamoja.* Kampala, Uganda: Centre for Basic Research.

Otterbein, K. 1995. 'More on the Nuer Conquest'. *Current Anthropology* vol. 36, no. 5: 821–23.

Pelican, M. 2004. 'Im Schatten der Schlachtviehmärkte: Milchwirtschaft bei den Mbororo in Nordwestkamerun', in G. Schlee (ed.), *Ethnizität und Markt: Zur ethnischen Struktur von Viehmärkten in Westafrika.* Cologne: Rüdiger Köppe Verlag, 131–58.

Perner, C. 1994. *The Anyuak, Living on Earth in the Sky: The Sphere of Spirituality,* vols 1 and 2. Basle: Halbig and Lichtenhehn.

Powles, J. M. 2000. 'Road 65: A Narrative Ethnography of an Angolan Refugee Settlement in Zambia'. DPhil thesis, University of Oxford.

Rake, A. (revised by M. Jennings) 2003. 'Uganda: Recent History', in Taylor and Francis Group, Europa Publications, *Africa South of the Sahara 2004.* London: Routledge, 1195–203.

Reno, W. 2002. 'Uganda's Politics of War and Debt Relief'. *Review of International Political Economy* vol. 9, no. 3: 415–35.

Rone, J. 2003. *Sudan, Oil, and Human Rights.* Washington, DC: Human Rights Watch.

Sahlins, 1961. 'The Segmentary Lineage: An Organisation of Predatory Expansion'. *American Anthropology* vol. 63: 322–44.

Sanderson, L.P. and G.M. Sanderson 1981. *Education, Religion and Politics in the Southern Sudan 1899–1964*. London: Ithaca.

Sato, R. 2002 'Evangelical Christianity and Ethnic Consciousness in Majangir. Remapping Ethiopia', in W. James, D.L. Donham, E. Kurimoto and A. Triulzi (eds), *Remapping Ethiopia: Socialism and After*. Oxford: James Currey, 185–97.

Schlee, G. 1979. *Das Glaubens- und Sozialsystem der Rendille: Kamelnomaden Nordkenias*. Berlin: Dietrich Reimer Verlag.

———— 1988. 'Camel Management Strategies and Attitudes towards Camels in the Horn', in J.C. Stone (ed.), *The Exploitation of Animals in Africa*. Aberdeen: Aberdeen University, African Studies Group, 143–54.

———— 1994a [1989]. *Identities on the Move: Clanship and Pastoralism in Northern Kenya*. Nairobi, Hamburg, Manchester: Gideon S. Were, LIT Verlag, Manchester University Press.

———— 1994b. 'Kuschitische Verwandtschaftssysteme in vergleichenden Perspektiven', in T. Geider and R. Kastenholz (eds), *Sprachen und Spracherzeugnisse in Afrika. Eine Sammlung philologischer Beiträge Wilhelm J.G. Möhlig zum 60. Geburtstag gewidmet*. Cologne: Rüdiger Köppe Verlag, 367–88.

———— 1994c. 'Ethnicity Emblems, Diacritical Features, Identity Markers: Some East African Examples', in D. Brokensha (ed.), *A River of Blessings*. New York: Syracuse University, 129–43.

———— 1997. 'Cross-cutting Ties and Interethnic Conflict: the Example of Gabbra, Oromo and Rendille', in K. Fukui, E. Kurimoto and M. Shigeta (eds), *Ethiopia in Broader Perspective: Papers of the 13th International Conference of Ethiopian Studies*, vol. 2. Kyoto: Shokado Book Sellers, 577–96.

———— 2000a. 'Identitätskonstruktionen und Parteinahme: Überlegungen zur Konflikttheorie'. *Sociologus* vol. 1: 64–89.

———— 2000b. 'Les Peuls du Nil', in Y. Diallo and G. Schlee (eds), *L'Ethnicité Peule dans des contextes nouveaux*. Paris: Karthala, 207–23.

———— 2002. 'Regularity in Chaos: the Politics of Difference in the Recent History of Somalia', in G. Schlee (ed.), *Imagined Differences: Hatred and the Construction of Identity*. Münster: LIT-Verlag, 251–80.

———— 2004a. 'Taking Sides and Constructing Identities: Reflections on Conflict Theory'. *Journal of the Royal Anthropological Instiute* vol. 10, no. 1: 135–56.

———— (ed.) 2004b. *Ethnizität und Markt: zur ethnischen Struktur von Viehmärkten in Westafrika* (with contributions by Jean Boutrais, Wei-Hsian Chi, Youssouf Diallo, Boris Nieswand, Michaela Pelican, Jens Rabbe). Cologne: Rüdiger Köppe Verlag.

———— 2008. *Ethiopian Diary 2001–2002 / Tagebuch Äthiopien 2001–2002*. (Online publication). Max Planck Institute for Social Anthropology. http://www.eth.mpg.de/subsites/schlee_tagebuch/index.html

Schlee, G. and M. Guichard 2007. 'Fulɓe und Usbeken im Vergleich', in *Max Planck Institute for Social Anthropology, Bericht 2007 (Abteilung I: Integration und Konflikt)*. Halle/Saale: Max Planck Institute for Social Anthropology, 11–53.

Schlee, G. and A.A. Shongolo (in preparation). *Islam and Ethnicity in Northern Kenya and Southern Ethiopia*.

Scott, G. 1990. 'A Resynthesis of the Primordial and Circumstantial Approaches to Ethnic Group Solidarity: Towards an Explanatory Model'. *Ethnic and Racial Studies* vol.13, no. 2: 147–71.

Sellen, D.W. and R. Mace 1999. 'A Phylogenetic Analysis of the Relationship between Sub-adult Mortality and Mode of Subsistence'. *Journal of Biosocial Science* vol. 31, no. 1: 1–16.

Shumet Sisagne 1986. 'The Economic Basis of Conflict Among the Nuer and Anuak Communities', in *Proceedings of the Third Annual Seminar of the Department of History*. Addis Ababa: Department of History, Addis Ababa University, 131–44.

Sobania, N. 1991. 'Feasts, Famines and Friends: Nineteenth Century Exchange and Ethnicity in the Eastern Lake Turkana Region' in J.G. Galaty and P. Bonte (eds), *Herders, Warriors, and Traders: Pastoralism in Africa*. Boulder, CO: Westview Press, 118–42.

Solomon, G. 1993. 'Nationalism and Ethnic Conflict in Ethiopia', in C. Young (ed.), *The Rising Tide of Cultural Pluralism: The Nation-state at Bay?*. Madison: University of Wisconsin Press, 138–57.

Spaulding, J. 1982. 'Slavery, Land Tenure and Social Class in the Northern Turkish Sudan'. *International Journal of African Historical Studies* vol. 15, no. 1: 1–20.

———— 1985. *The Heroic Age in Sinnar*. East Lansing: Michigan State University.

Spear, T. 1993. 'Introduction', in T. Spear and R. Waller (eds), *Being Maasai: Ethnicity and Identity in East Africa*. Oxford: James Currey, 1–18.

Spencer, P. 1965. *The Samburu: a Study of Gerontocracy in a Nomadic Tribe*. Berkeley: University of California Press.

———— 1973. *Nomads in Alliance*. London: Oxford University Press.

Sundal, M., S.J. Gray and H.A. Akol 2001. 'Child Mortality among Karimojong Agropastoralists of Northeast Uganda'. *American Journal of Physical Anthropology Supplement* vol. 32: 146.

Swift, J. 1996. 'Desertification', in M. Leach and R. Mearns (eds), *The Lie of the Land: Challenging Received Wisdom on the African Environment*. London: The International African Institute, 73–90.

Thomas, E.M. 1965. *Warrior Herdsmen*. New York: Knopf.

Thompson, R. 1989. *Theories of Ethnicity: A Critical Appraisal*. New York: Greenwood Press.

Tornay, S. 2001. *Les Fusils jaunes: Générations et politique en pays Nyangatom (Éthiopie)*. Nanterre: Société d' Ethnologie.

Turton, D. 1997. 'War and Ethnicity: Global Connections and Local Violence in North-east Africa and Former Yugoslavia'. *Oxford Development Studies* vol. 25, no. 1: 77–94.

UNDP 2003. *Human Development Report 2003*. Oxford: Oxford University Press.

UNHCR 1994. 'Briefing: Population Census in Kakuma Camp', 22 December 1994. Nairobi: UNHCR Nairobi Office.

———— 2000. 'Population Breakdown by Nationality as of 13th September 2000'. Kakuma: UNHCR Sub-office.

van Acker, F. 2004. 'Uganda and the Lord's Resistance Army: the New Order No One Ordered'. *African Affairs* vol. 103, no. 412: 335–57.

van Santen, J. 2000. '"Garder un bétail, c'est aussi du travail": Les relations entre pasteurs peuls et agriculteurs du centre du Bénin et du Nord-Cameroun', in Y. Diallo and G. Schlee (eds), *L'Ethnicité Peule dans des contextes nouveaux*. Paris: Karthala, 129–59.

VerEecke C. 1988. 'Pulaaku: Adamawa Fulbe Identity and its Transformations'. PhD dissertation, University of Pennsylvania.

Verney, P. 1999. *Raising the Stakes: Oil and Conflict in Sudan*. Hebden Bridge: Sudan Update.

Weber, M. 2001. *The Protestant Ethic and the Spirit of Capitalism*. London: Routledge.

Wiebusch, B., S.J. Gray and H.A. Akol 2001. 'Child Growth among Karimojong Agropastoralists of Northeast Uganda'. *American Journal of Physical Anthropology Supplement* vol. 32: 164–65.

Wieland, C. 2001. 'Ethnic Conflict Undressed: Patterns of Contrast, Interest of Elites, and Clientelism of Foreign Powers in Comparative Perspective – Bosnia, India, and Pakistan'. *Nationalities Papers* vol. 29, no. 2: 207–41.

Wilson, J.G. 1985. 'Resettlement in Karamoja', in C.P. Dodge and P.D. Wiebe (eds), *Crisis in Uganda: The Breakdown of Health Services*. Oxford: Pergamon Press, 163–70.

Wilson, T.M. and H. Donnan (eds) 1998. *Border Identities: Nation and State at International Frontiers*. Cambridge: Cambridge University Press.

Yacob, A. et al. 2000. 'Traditional Conflict Prevention and Management. Cases for Local Capacity for Peace in Gambella and Afar'. Research Report submitted to the Ethiopian Committee for the Red Cross, Addis Ababa.

Young, J. 1999. 'Along Ethiopia's Western Frontier: Gambela and Benishangul in Transition'. *Journal of Modern African Studies* vol. 37, no. 2: 321–46.

——— 2007. 'The White Army: An Introduction and Overview'. *Sudan Human Security Baseline Assessment Project, Working Paper 5*. Geneva: Small Arms Survey.

Notes on Contributors

Al-Amin Abu-Manga is Professor at the Institute of African and Asian Studies at the University of Khartoum. He belongs to the Fallata community of Maiurno. His father can be described as the most successful member of the third generation of lorry drivers in Maiurno (the subject of his chapter in Volume II), owning in the mid-1970s up to three lorries and a Land-Rover.

Janice Boddy is Professor of Anthropology at the University of Toronto. She works on gender, identity and the body in the Middle East and North-East and Eastern Africa. As well as contributing articles to many books, her monographs include *Civilizing Women: British Crusaders in Colonial Sudan* (Princeton, 2007) and *Wombs and Alien Spirits: Women, Men and the Zar Cult in Northern Sudan* (Wisconsin, 1989).

Dereje Feyissa is currently a Research Fellow at the Max Planck Institute for Social Anthropology. He has carried out research on the conflict situation in the Gambella region and his recent publications include 'Land and the Politics of Identity, the Case of Anuak-Nuer Relations in the Gambela Region, Western Ethiopia' (in Evers and Spierenburg, eds, 2005, *Competing Jurisdictions: Settling Land Claims in Africa*) and 'The Federal Experience of the Gambella Regional State' (in Turton, ed., 2006, *Ethnic Federalism in Ethiopia: Prospects and Challenges*).

Christiane Falge is a former PhD student at the Max Planck Institute for Social Anthropology, and has done extensive research in the Horn of Africa. Currently she is research coordinator of a study group on migration and health based in the Centre of Social Science Research (CSSR), University of Bremen. She continues to be affiliated to the Max Planck Institute, and is working on a GTZ project as a consultant for conflict resolution in Ethiopia with support from Günther Schlee. As part of a research team led by Eisei Kurimoto, Falge is continuing to observe the events after the peace agreement in Sudan, and is focusing on returning Nuer refugees and the way they try to accommodate problems after returning home.

Sandra Gray is Associate Professor of Anthropology at the University of Kansas, where she also serves as an adviser for the Human Biology Programme. She currently is preparing a monograph on the Karimojong conflict, Forgotten: Life and Deaths in a Little War, which is based on research carried out in 2004.

Sharon Elaine Hutchinson is Professor of Anthropology and Director of the African Studies Programme at the University of Madison-Wisconsin. Her publications include the highly acclaimed *Nuer Dilemmas: Coping with Money, War, and the State* (Berkeley: University of California Press, 1996).

Wendy James is Emeritus Professor of Social Anthropology at the University of Oxford and a Fellow of St Cross College. Her recent books include *War and Survival in Sudan's Frontierlands: Voices from the Blue Nile* (Oxford: Oxford University Press, 2007), *The Ceremonial Animal: A New Portrait of Anthropology* (Oxford: Oxford University Press, 2003), *Remapping Ethiopia: Socialism and After* (ed. with D. Donham, E. Kurimoto and A. Triulzi, Oxford: James Currey, 2002), and *Anthropologists in a Wider World: Essays on Field Research* (ed. with P. Dresch and D. Parkin, New York and Oxford: Berghahn, 2000).

Douglas H. Johnson is an historian and publisher based in Oxford, UK. His publications include *Governing the Nuer: Documents by Percy Coriat on Nuer History and Ethnography, 1922–1931* (Oxford: JASO, 1993), *Nuer Prophets: A History of Prophecy from the Upper Nile in the 19th and 20th Centuries* (Oxford: Oxford University Press, 1994), *The Root Causes of Sudan's Civil Wars* (Oxford: James Currey, 2003).

Eisei Kurimoto is Professor of Anthropology at the University of Osaka. He has carried out sustained research in Sudan and Ethiopia, with particular reference to the Pari and the Anywaa. He has published (with Katsuyoshi Fukui and Masayoshi Shigeta) three volumes *Ethiopia in Broader Perspective*, Kyoto: Shokado 1997.

Günther Schlee is a Director at the Max Planck Institute for Social Anthropology at Halle. Until 1999 he was a Professor at Bielefeld. His habilitation thesis (Bayreuth, 1986) has been published as *Identities on the Move: Clanship and Pastoralism in Northern Kenya* in the series of the International African Institute (1989). In his department, Integration and Conflict, he directs research in Africa, Central Asia and Europe.

Elizabeth E. Watson is a Lecturer in the Department of Geography at the University of Cambridge. Her first degree is in anthropology and her second in geography. Her research focuses on environment and development issues, mainly in Eastern Africa, and most of her work in Ethiopia has been among the Konso. Recent publications include: 'Local Community, Legitimacy, and Cultural Authenticity in Postconflict Natural Resource Management: Ethiopia and Mozambique', *Environment and Planning D: Society and Space*, 2006 (with R. Black), and 'Making a Living in the Post-Socialist Periphery: Struggles between Farmers and Traders in Konso, Ethiopia', *Africa*, 2006.

Index

www.ingramcontent.com/pod-product-compliance
Lightning Source LLC
Chambersburg PA
CBHW060030030426
42334CB00019B/2265